Tutorials, Schools, and Workshops in the Mathematical Sciences

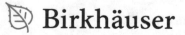 Birkhäuser

This series will serve as a resource for the publication of results and developments presented at summer or winter schools, workshops, tutorials, and seminars. Written in an informal and accessible style, they present important and emerging topics in scientific research for PhD students and researchers. Filling a gap between traditional lecture notes, proceedings, and standard textbooks, the titles included in TSWMS present material from the forefront of research.

More information about this series at http://www.springer.com/series/15641

Maria Ulan • Eivind Schneider

Editors

Differential Geometry, Differential Equations, and Mathematical Physics

The Wisła 19 Summer School

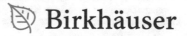 Birkhäuser

Editors
Maria Ulan
Baltic Institute of Mathematics
Warszawa, Poland

Eivind Schneider
Baltic Institute of Mathematics
Warszawa, Poland

ISSN 2522-0969   ISSN 2522-0977   (electronic)
Tutorials, Schools, and Workshops in the Mathematical Sciences
ISBN 978-3-030-63255-7   ISBN 978-3-030-63253-3   (eBook)
https://doi.org/10.1007/978-3-030-63253-3

Mathematics Subject Classification: 53-01, 76M60, 35A30, 35Q35, 53B50, 53D17

This book is published under the imprint Birkhäuser, www.birkhauser-science.com by the registered
company Springer Nature Switzerland AG
The registered company address is: Gewerbestrasse 11, 6330 Cham, Switzerland

Preface

The Summer School Wisla 19: Differential Geometry, Differential Equations, and Mathematical Physics was organized by the Baltic Institute of Mathematics and took place on August 19–29, 2019, in the beautiful mountain region of Wisła, Poland.

The school was devoted to symplectic and Poisson geometry, tractor calculus, and integration of ordinary differential equations. There were three series of main lectures, given by Vladimir Roubtsov, Jan Slovák, and Valentin Lychagin, respectively:

- Introduction to Symplectic and Poisson Geometry of Integrable Systems
- Tractor Calculi
- How to Integrate Differential Equations

The corresponding lecture notes make up the first three chapters of this book. It is our pleasure to share these inspiring lectures, given by experts in their fields, with an audience greatly exceeding that of those who were fortunate to be in Wisła when they were given.

The subsequent chapters continue the tradition of the previous three by introducing the reader to different topics at the intersection of differential geometry, differential equations, and mathematical physics. They are written in a pedagogical style while simultaneously bringing to attention recent advances made by their authors. The book is aimed at students and researchers who are looking for a concise introduction to the topics covered here. Since all six chapters are written from a geometric perspective, the reader is expected to have some basic knowledge of differential geometry. Below is a summary of each chapter:

Poisson and Symplectic Structures, Hamiltonian Action, Momentum and Reduction is written in a concise form and gives a brief review of well-known material. It covers Poisson and symplectic structures, group actions and orbits, moment maps, and Poisson and Hamiltonian actions. In the end, phase space reduction and Poisson–Lie structures are discussed.

Notes on Tractor Calculi presents an elementary introduction to tractors based on classical examples, together with glimpses toward modern invariant differential calculus related to a vast class of Cartan geometries, the so-called parabolic geometries. The exposition aims at a quick understanding of basic principles, omitting many proofs or at least their details. Some knowledge in representation theory is assumed.

Symmetries and Integrals is an invitation to the world of symmetries and differential equations. It starts with an introduction to distributions and moves quickly to the challenge of finding explicitly, in quadratures, integral manifolds of completely integrable distributions. While these lecture notes rely heavily on the insight that a geometric understanding of distributions and symmetries gives, they simultaneously focus on the practical aspects of finding, and writing down, exact solutions of differential equations.

Finite Dimensional Dynamics of Evolutionary Equations with Maple uses geometric methods to find exact solutions to partial differential equations appearing in physics, mathematical biology, and mathematical finance. The authors emphasize the computational aspect and provide detailed Maple code. This makes the chapter an excellent introduction to the subject of finite dimensional dynamics by allowing the reader to start using the methods instantaneously.

Critical Phenomena in Darcy and Euler Flows of Real Gases is a survey article on recent results obtained by its authors concerning gas flows through porous media and flows of inviscid gases. Thermodynamics is formulated in terms of contact and symplectic geometry, and the link to measurement theory is emphasized. The methods provided by the geometric formulation of thermodynamics are applied to the analysis of various models of real gases, and special attention is paid to phase transitions. The thermodynamics of the gases under consideration emerges along the gas flow. Explicit methods for finding solutions to the Dirichlet filtration problem and Euler flows are provided. In particular, the locations for different phases of the medium are found.

Differential Invariants for Flows of Fluids and Gases summarizes several of the authors' earlier results while simultaneously improving them by taking into account additional structure on the thermodynamic states. Similar to the previous chapter, it relies on the symplectic and contact-geometric formulation of thermodynamics. After adding thermodynamic equations of state to the Navier–Stokes and Euler equations, the authors compute point symmetries of the equations. The symmetry Lie algebra depends, in general, on the equations of state, and an analysis of possible symmetry Lie algebras is made before the corresponding differential invariants are computed.

We hope that this book will function as a good entry point to the topics covered here and that it will aid the reader with motivation and competence to dive deeper into the world of differential geometry and mathematical physics.

Warszawa, Poland   Maria Ulan

Warszawa, Poland into Hradec Králové, The Czech Republic   Eivind Schneider
July 2020

Sunset over the mountains surrounding Wisła, August 2019

Acknowledgments

It would have been difficult to organize a school like this without the help and effort from the community.

Baltic Institute of Mathematics would like to thank Prof. Valentin Lychagin and Prof. Olav Arnfinn Laudal for their kindliness, expertise, and continuous support. Extra acknowledgments go to Prof. Jerzy Buzek, MEP, for the honorary patronage of the Futurum 2020 Initiative that helps to organize and provide excellent meeting facilities during the school.

The organizers of the school would like to thank all participants of the school for a great, friendly atmosphere and strong motivation to knowledge sharing, and special thanks go to Charli Benjamin for arranging extra social activity. The speakers of the main lectures and the authors deserve a special thank you for their contributions to the success of the school.

The editors would like to thank the reviewers for their valuable comments and suggestions on preliminary drafts of the book, and the Birkhäuser Mathematics, Springer Nature crew, especially Chris Tominich, for their assistance.

Contents

1   **Poisson and Symplectic Structures, Hamiltonian Action, Momentum and Reduction**   1
Vladimir Roubtsov and Denys Dutykh

2   **Notes on Tractor Calculi**   31
Jan Slovák and Radek Suchánek

3   **Symmetries and Integrals**   73
Valentin V. Lychagin

4   **Finite Dimensional Dynamics of Evolutionary Equations with Maple**   123
Alexei G. Kushner and Ruslan I. Matviichuk

5   **Critical Phenomena in Darcy and Euler Flows of Real Gases**   151
Valentin V. Lychagin and Mikhail D. Roop

6   **Differential Invariants for Flows of Fluids and Gases**   187
Anna Duyunova, Valentin V. Lychagin, and Sergey Tychkov

Contributors

Denys Dutykh University Grenoble Alpes, University Savoie Mont Blanc, CNRS, LAMA, Chambéry, France
LAMA, UMR 5127 CNRS, Université Savoie Mont Blanc, Le Bourget-du-Lac Cedex, France

Anna Duyunova V. A. Trapeznikov Institute of Control Sciences of Russian Academy of Sciences, Moscow, Russia

Alexei G. Kushner Lomonosov Moscow State University, GSP-2, Moscow, Russia
Moscow Pedagogical State University, Moscow, Russia

Valentin V. Lychagin V. A. Trapeznikov Institute of Control Sciences of Russian Academy of Sciences, Moscow, Russia

Ruslan I. Matviichuk Lomonosov Moscow State University, GSP-2, Moscow, Russia

Mikhail D. Roop Institute of Control Sciences of Russian Academy of Sciences, Moscow, Russia
Lomonosov Moscow State University, Moscow, Russia

Vladimir Roubtsov LAREMA UMR 6093, CNRS & Université d'Angers, Angers Cedex 01, France
IGAP (Institute of Geometry and Physics), Trieste, Italy

Jan Slovák Department of Mathematics and Statistics, Masaryk University, Brno, Czech Republic

Radek Suchánek Department of Mathematics and Statistics, Masaryk University, Brno, Czech Republic
LAREMA UMR 6093, CNRS & Université d'Angers, Angers Cedex 01, France

Sergey Tychkov V. A. Trapeznikov Institute of Control Sciences of Russian Academy of Sciences, Moscow, Russia

Chapter 1
Poisson and Symplectic Structures, Hamiltonian Action, Momentum and Reduction

Vladimir Roubtsov and Denys Dutykh

1.1 Introduction

These lectures have been delivered by the first author in the Summer School *"Differential Geometry, Differential Equations, and Mathematical Physics"* at Wisła, Poland from 19th to 29th of August 2019. The second author took the notes of these lectures.

As the title suggests, the material covered here includes the Poisson and symplectic structures (Poisson manifolds, Poisson bi-vectors, and Poisson brackets), group actions and orbits (infinitesimal action, stabilizers, and adjoint representations), moment maps, Poisson and Hamiltonian actions. Finally, the phase space reduction is also discussed. The Poisson structures are a particular instance of Jacobi structures introduced by A. Lichnerowicz back in 1977 [7]. Several capital contributions to this field were made by A. Weinstein, see e.g. [13].

The text below does not pretend to provide any new scientific results. However, we believe that this point of view and exposition will be of some interest to our readers. As other general (and excellent) references on this topic include:

- R. Abraham and J.E. Marsden. *Foundations of Mechanics*, second edn., Addison–Wesley Publishing Company, Redwood City, CA, 1987 [1]

V. Roubtsov
LAREMA UMR 6093, CNRS and Université d'Angers, Angers, France
IGAP (Institute of Geometry and Physics), Trieste, Italy
e-mail: vladimir.roubtsov@univ-angers.fr

D. Dutykh (✉)
University Grenoble Alpes, University Savoie Mont Blanc, CNRS, LAMA, Chambéry, France

LAMA, UMR 5127 CNRS, Université Savoie Mont Blanc, Le Bourget-du-Lac, France
e-mail: denys.dutykh@univ-smb.fr

© The Author(s), under exclusive license to Springer Nature Switzerland AG 2021
M. Ulan, E. Schneider (eds.), *Differential Geometry, Differential Equations, and Mathematical Physics*, Tutorials, Schools, and Workshops in the Mathematical Sciences, https://doi.org/10.1007/978-3-030-63253-3_1

- V.I. Arnold. *Mathematical methods of classical mechanics*, second edn., Springer, New York, 1997 [2]
- A.M. Vinogradov and B.A. Kupershmidt. *The structures of Hamiltonian mechanics*, Russ. Math. Surv., **32**(4), 177–243, 1977 [12]

We can mention also another set of recent lecture notes on the symplectic and contact geometries [11]. We mention also the classical review of this topic [14].

Our presentation remains at quite elementary level of exposition. We restricted deliberately ourselves to the presentation of basic notions and the state of the art as it was in 1990–2000. Nevertheless, we hope that motivated students will be inspired to find more advanced and modern material which is inevitably based on these elementary notes.

In the following text each Section corresponds to a separate lecture. This text is organized as follows. The Poisson and symplectic structures are presented in Sect. 1.2. The group actions and orbits are introduced in Sect. 1.3. The moment map, Poisson and Hamiltonian actions are described in Sect. 1.4. Finally, the manuscript is ended with the description of the phase space reduction in Sect. 1.5. The very last Sect. 1.6 is a brief (and essentially incomplete) introduction to Poisson–Lie structures and some related notions. An excellent account of the last topic can be found in the survey paper by Y. Kossmann-Schwarzbach (1997) [6].

1.2 Poisson and Symplectic Structures

Hamiltonian systems are usually introduced in the context of the symplectic geometry [5, 10]. However, the use of Poisson geometry emphasizes the Lie algebra structure, which underlies the Hamiltonian mechanics.

1.2.1 Poisson Manifolds

Let $\mathbb{M}$ be a smooth manifold with a bracket

$$\{-, -\}:\ C^\infty(\mathbb{M}) \times C^\infty(\mathbb{M}) \longmapsto C^\infty(\mathbb{M}),$$

which verifies the following properties:

- Bi-linearity $\{-, -\}$ is real-bilinear.
- Anti-symmetry $\{F, G\} = -\{G, F\}$.
- Jacobi identity $\{\{F, G\}, H\} + \{\{G, H\}, F\} + \{\{H, F\}, G\} = 0$.
- Leibniz identity $\{FG, H\} = F\{G, H\} + \{F, H\}G$.

Then, the bracket $\{-, -\}$ is a Poisson bracket and the pair $\left(\mathbb{M}; \{-, -\}\right)$ will be called a Poisson manifold. A Poisson algebra is defined as the following pair

$(C^\infty(\mathbb{M}; \mathbb{R}); \{-, -\})$. Thanks to the first three properties of the Poisson bracket, it is not difficult to see that $(C^\infty(\mathbb{M}; \mathbb{R}); \{-, -\})$ is also a Lie algebra. The last property of the Poisson bracket (i.e. the Leibniz identity) implies that it is also a *derivative* in each of its arguments.

Let $(\mathbb{M}; \{-, -\})$ be a Poisson manifold and $H \in C^\infty(\mathbb{M}; \mathbb{R})$, then there exists a unique vector field $\mathcal{X}_H$ such that

$$\mathcal{X}_H(G) = \{G, H\}, \qquad \forall G \in C^\infty(\mathbb{M}; \mathbb{R}).$$

The vector field $\mathcal{X}_H$ is called the Hamiltonian vector field with respect to the Poisson structure with H being the Hamiltonian function. Let $\mathfrak{X}(\mathbb{M})$ denote the space of all vector fields on $\mathbb{M}$. Then, the just constructed mapping $C^\infty(\mathbb{M}; \mathbb{R}) \longrightarrow \mathfrak{X}(\mathbb{M})$ is a Lie algebra morphism, i.e. $\mathcal{X}_{\{F,G\}} = [\mathcal{X}_F, \mathcal{X}_G]$.

Definition 1.1 A Casimir function on a Poisson manifold $(\mathbb{M}; \{-, -\})$ is a function $F \in C^\infty(\mathbb{M}; \mathbb{R})$ such that for all $G \in C^\infty(\mathbb{M}; \mathbb{R})$ one has

$$\{F, G\} = 0, \qquad \forall G \in C^\infty(\mathbb{M}; \mathbb{R}).$$

1.2.2 Poisson Bi-vector

If $(\mathbb{M}; \{-, -\})$ is a Poisson manifold, then there exists a contravariant anti-symmetric two-tensor $\pi \in \Lambda^2(\mathbb{TM})$ or equivalently

$$\pi : \mathbb{T}^*\mathbb{M} \times \mathbb{T}^*\mathbb{M} \longrightarrow \mathbb{R}$$

such that

$$\langle \pi, dF \wedge dG \rangle(z) = \pi(z)(dF(z); dG(z)) = \{F, G\}(z).$$

In local coordinates $(z_1, z_2, \ldots, z_n)$ we have the following expression for the Poisson bracket:

$$\{F, G\} = \sum_{i, j} \pi^{ij} \frac{\partial F}{\partial z_i} \frac{\partial G}{\partial z_j},$$

where $\pi^{ij} \overset{\text{def}}{=} \{z_i, z_j\}$ are called the elements of the *structure matrix* or the Poisson bi-vector of the underlying Poisson structure. For the vector field we have the corresponding expression in coordinates:

$$\mathcal{X}_H = \sum_{i, j} \pi^{ij} \frac{\partial H}{\partial z_i} \frac{\partial}{\partial z_j}, \qquad \text{or} \qquad \mathcal{X}_H^j = \sum_i \pi^{ij} \frac{\partial H}{\partial z_i}.$$

1.2.2.1 Hamilton Map and Jacobi Identity

Let $\pi = (\pi^{ij})$ be a Poisson bi-vector on $\mathbb{M}$, then there exists a $C^{\infty}(\mathbb{M}; \mathbb{R})$−linear map $\pi^{\sharp} : \mathbb{T}^{*}\mathbb{M} \longmapsto \mathbb{T}\mathbb{M}$ given by

$$\pi^{\sharp}(\alpha) \lrcorner \beta = \{\pi, \alpha \wedge \beta\}(z) = \pi(z)\big(\alpha(z), \beta(z)\big),$$

where $\lrcorner$ denotes the usual interior product or the substitution of a vector field into a form.

If $\alpha = \mathrm{d}f$ for some $f \in C^{\infty}(\mathbb{M}; \mathbb{R})$, then $\pi^{\sharp}(\mathrm{d}H) = \mathcal{X}_H$. It is not difficult to see how the map $\pi^{\sharp}$ acts on the basis elements of co-vectors:

$$\pi^{\sharp}(\mathrm{d}z_i) = \{z_i, z_j\}\frac{\partial}{\partial z_j}.$$

Finally, we also have the following Jacobi identity:

$$\pi^{il}\frac{\partial \pi^{jk}}{\partial z_l} + \pi^{jl}\frac{\partial \pi^{ki}}{\partial z_l} + \pi^{kl}\frac{\partial \pi^{ij}}{\partial z_l} = 0. \tag{1.1}$$

1.2.3 Symplectic Structures on Manifolds

Definition 1.2 A *symplectic form* on a real manifold $\mathbb{M}$ is a non-degenerate closed 2−form $\omega \in \Omega^2(\mathbb{M}) \overset{\text{def}}{:=} \Lambda^2(\mathbb{T}^{*}\mathbb{M})$. Such a manifold is called a *symplectic manifold* and it is denoted by a couple $(\mathbb{M}; \omega)$.

Let $(\mathbb{M}; \{-, -\})$ be a Poisson manifold with non-degenerate Poisson structure bi-vector (π^{ij}) and the Hamiltonian isomorphism $\pi^{\sharp}$ such that $\pi^{\sharp}(\alpha) = \mathcal{X}$. Then, there is the inverse map $(\pi^{\sharp})^{-1} : \mathbb{T}\mathbb{M} \longrightarrow \mathbb{T}^{*}\mathbb{M}$ is defined by the following relation:

$$\mathcal{Y} \lrcorner (\pi^{\sharp})^{-1}(\mathcal{X}) = \alpha(\mathcal{Y}).$$

Moreover, the inverse operator $(\pi^{\sharp})^{-1}$ defines a 2−form ω_{π} as follows:

$$\omega_{\pi}(\mathcal{X}, \mathcal{Y}) = \big\langle(\pi^{\sharp})^{-1}(\mathcal{X}), \mathcal{Y}\big\rangle.$$

1.2.3.1 Darboux Theorem and Hamiltonian Vector Fields

The following Lemma describes some important properties of the just defined form ω_{π} along with the underlying manifold $\mathbb{M}$:

Lemma 1.1 *The real manifold* $\mathbb{M}$ *always has an even dimension. The form* ω_π *is a symplectic 2-form on* $\mathbb{M}$. *The Jacobi identity* (1.1) *is equivalent to* $d\omega_\pi = 0$.

Proof Left to the reader as an exercise.   □

Theorem 1.1 (G. Darboux) *Let* $(\mathbb{M}; \omega)$ *be a symplectic manifold. There exists a local coordinate system* $(q_1, q_2, \ldots, q_n, p_1, p_2, \ldots, p_n) \overset{\text{def}}{=:} (\mathbf{q}, \mathbf{p})$ *such that* $\omega = \sum_{i=1}^{n} d q_i \wedge d p_i$. *Such coordinates are called* canonical *or* Darboux *coordinates.*

Lemma 1.2 *Let* $(\mathbb{M}; \omega)$ *be a symplectic manifold and* $H \in C^\infty(\mathbb{M}; \mathbb{R})$ *the Hamiltonian function. Then, there is a unique vector field* $\mathcal{X}_H$ *(i.e. the Hamiltonian vector field associated with the Hamiltonian* H) *on* $\mathbb{M}$ *such that*

$$\mathcal{X}_H \lrcorner \omega = dH.$$

The Hamiltonian vector field $\mathcal{X}_H$ *can be written in the canonical coordinates* $(\mathbf{q}, \mathbf{p})$ *on* $\mathbb{M}$ *as*

$$\mathcal{X}_H = \sum_i \left(\frac{\partial H}{\partial p_i} \frac{\partial}{\partial q_i} - \frac{\partial H}{\partial q_i} \frac{\partial}{\partial p_i} \right).$$

The Poisson bracket in these coordinates looks like

$$\{F, G\} = \mathcal{X}_F(G) = \sum_{i,j} \left(\frac{\partial F}{\partial p_i} \frac{\partial G}{\partial q_i} - \frac{\partial F}{\partial q_i} \frac{\partial G}{\partial p_i} \right).$$

Proof Left to the reader as an exercise.   □

1.2.3.2 Example

Let $\mathbb{S}^2 \overset{\text{def}}{:=} \{ (x, y, z) \in \mathbb{R}^3 \mid x^2 + y^2 + z^2 = 1 \}$ be the 2-sphere, which can be naturally injected in $\mathbb{R}^3 : \iota : \mathbb{S}^2 \hookrightarrow \mathbb{R}^3$. The 2-form $\bar{\omega} \in \Lambda^2(\mathbb{R}^3)$ is given by

$$\bar{\omega} = x \, dy \wedge dz + y \, dz \wedge dx + z \, dx \wedge dy$$

and $\omega \in \Lambda^2(\mathbb{S}^2)$ is defined as $\omega = \iota^*(\bar{\omega})$.

Lemma 1.3 *The form* ω *gives a symplectic structure on* $\mathbb{S}^2$, *i.e.* $d\omega = 0$ *and this 2-form is non-degenerate on* $\mathbb{S}^2$.

Proof First of all, we observe that the closedness of the 2-form ω is a straightforward conclusion in view of

$$d\omega = d\big(\imath^*(\bar{\omega})\big) = \imath^*\big(d\bar{\omega}\big) = 0,$$

since $\Lambda^3(\mathbb{S}^2) = 0$. To check that it is non-degenerate, we make a choice of the following charts:

$$\phi_N : \mathbb{S}^2 \setminus \{N\} \longrightarrow \mathbb{R}^2,$$

$$(x, y, z) \longmapsto \left(\frac{x}{1-z}, \frac{y}{1-z} \right),$$

$$\phi_S : \mathbb{S}^2 \setminus \{S\} \longrightarrow \mathbb{R}^2,$$

$$(x, y, z) \longmapsto \left(\frac{x}{1+z}, \frac{y}{1+z} \right).$$

It is not difficult to see that their inverses are given by the following maps:

$$\phi_N^{-1} : (u, v) \longmapsto \left(\frac{2u}{1+u^2+v^2}, \frac{2v}{1+u^2+v^2}, \frac{u^2+v^2-1}{1+u^2+v^2} \right),$$

$$\phi_S^{-1} : (u, v) \longmapsto \left(\frac{2u}{1+u^2+v^2}, \frac{2v}{1+u^2+v^2}, -\frac{u^2+v^2-1}{1+u^2+v^2} \right).$$

In both coordinate charts (u, v) induced by $\phi_{N,S}$ we obtain

$$\imath^*(\bar{\omega}) = (x\circ\imath)\,d(y\circ\imath)\wedge d(z\circ\imath) + (y\circ\imath)\,d(z\circ\imath)\wedge d(x\circ\imath) + (z\circ\imath)\,d(x\circ\imath)\wedge d(y\circ\imath)$$

$$= -\frac{4}{1+u^2+v^2}\,du\wedge dv \neq 0.$$

1.2.4   Co-tangent Bundle: Liouville Form

Let $\mathbb{M}$ be a smooth n−dimensional manifold and $\varpi : T^*\mathbb{M} \longrightarrow \mathbb{M}$ is the projection map whose differential map is $T_\varpi : TT^*\mathbb{M} \longrightarrow T\mathbb{M}$.

Definition 1.3 A differential 1−form ρ on $T^*\mathbb{M}$, which is defined by $\sigma_\rho : T^*\mathbb{M} \longrightarrow T^*T^*\mathbb{M}$ as follows:

$$\langle \sigma_\rho, \mathcal{X} \rangle = \langle \rho, T_{\varpi_\rho}(\mathcal{X}) \rangle, \qquad \mathcal{X} \in T_{\varpi_\rho}(T^*\mathbb{M})$$

is called the Liouville (or *action*) form.

If $(\mathbf{q}, \mathbf{p})$ is a local coordinate system on $T^*\mathbb{M}$, then the form ρ can be written as

$$\rho = \mathbf{p}\,d\mathbf{q} = \sum_{i=1}^{n} p_i\,dq^i\,.$$

The 2−form $\omega \in \Omega^2(\mathbb{T}^*M)$, $\omega = d\rho = d\mathbf{p} \wedge d\mathbf{q} = \sum_{i=1}^{n} dp_i \wedge dq^i$ is the *canonical* symplectic form.

1.2.5 Non-Symplectic Poisson Structures

Let $\mathfrak{g}$ be a real finite-dimensional Lie algebra and $\mathfrak{g}^*$ be its dual (as a vector space). If we suppose that $\dim \mathfrak{g} = n$, then $\mathfrak{g}^*$ is isomorphic as a real smooth manifold to $\mathbb{R}^n$.

Theorem 1.2 *There exists a (non-symplectic) Poisson structure on* $\mathfrak{g}^*$.

1.2.6 Poisson Brackets on Dual of a Lie Algebra

A bracket can be defined on $C^\infty(\mathfrak{g}^*)$ by the identification:

$$\mathfrak{g} \simeq \mathfrak{g}^{**} \equiv C^\infty_{\mathrm{lin}}(\mathfrak{g}^*) \subset C^\infty(\mathfrak{g}^*), \qquad \mathcal{X} \longmapsto F_{\mathcal{X}}$$

$$F_{\mathcal{X}}(\xi) = \langle \xi, \mathcal{X} \rangle = \xi(\mathcal{X})\,.$$

One has $\left\{ F_{\mathcal{X}}, F_{\mathcal{Y}} \right\} = F_{[\mathcal{X},\mathcal{Y}]}$. Let $\{e_i\}_{i=1}^{n}$ be a base of $\mathfrak{g}$ and $\left\{ C_{ij}^k \right\}$ be structure constants of the Lie algebra $\mathfrak{g}$, i.e.

$$[e_i, e_j] = \sum_{k=1}^{n} C_{ij}^k\, e_k\,.$$

Let $\{F_i\}_{i=1}^{n}$ be a dual basis of $\mathfrak{g}^*$ and $\{\mathcal{X}_i\}_{i=1}^{n}$ be the coordinate system of $\mathfrak{g}^*$, i.e.

$$\xi = \mathcal{X}_i(\xi)F_i\,, \qquad \mathcal{X}_i(F_j) = \delta_{ij}\,, \qquad \mathcal{X}_i = F_{e_i}\,.$$

Now, it is not difficult to see that

$$\left\{ \mathcal{X}_i, \mathcal{X}_j \right\} = F_{[e_i, e_j]} = \sum_{k} C_{ij}^k\, F_{e_k} = \sum_{k} C_{ij}^k\, \mathcal{X}_k\,.$$

Finally, we can write the coordinate expression of the Poisson bracket on the dual of a Lie algebra:

$$\{F, G\} = \sum_{i, j, k} C_{ij}^{k} \frac{\partial F}{\partial \mathcal{X}_i} \frac{\partial G}{\partial \mathcal{X}_j} \mathcal{X}_k.$$

1.2.6.1 Definition via Gradient Operator

We define the gradient operator

$$\nabla : C^{\infty}(\mathfrak{g}^*) \longrightarrow C^{\infty}(\mathfrak{g}^*)$$

as follows:

$$\langle \eta, \nabla F(\xi) \rangle \overset{\text{def}}{:=} \frac{d}{dt} F(\xi + t\eta)\Big|_{t=0}, \qquad \forall F \in C^{\infty}(\mathfrak{g}^*).$$

In coordinates one simply has

$$\nabla F = \frac{\partial F}{\partial \mathcal{X}_i} e_i.$$

Using the gradient operator, the bracket is defined as

$$\{F, G\}(\xi) \overset{\text{def}}{:=} \langle \xi, [\nabla F(\xi), \nabla G(\xi)] \rangle$$

and in coordinates we obtain

$$\left\{ \mathcal{X}_i, \mathcal{X}_j \right\}(F_k) = \left\langle F_k, [\nabla \mathcal{X}_i(F_k), \nabla \mathcal{X}_j(F_k)] \right\rangle = \left\langle F_k, [e_i, e_j] \right\rangle = C_{ij}^{k}.$$

1.2.6.2 Definition via Canonical Structure on $\mathbb{T}^*(G)$

Let $\mathfrak{g}$ be a Lie algebra, then there exists a unique (connected and simply connected up to an isomorphism) Lie group G such that $\mathbb{T}_e G \simeq \mathfrak{g}$. Let $L_g \in \mathbf{Aut}(G)$ be the translation by g, i.e. $\forall h \in G$:

$$L_g : G \xrightarrow{\simeq} G$$
$$h \longmapsto gh.$$

Then, we define

$$\lambda_g(h) \overset{\text{def}}{:=} \left(\mathbb{T}L_g(h)\right)^* : \mathbb{T}_{gh}^*(G) \longrightarrow \mathbb{T}_h^*(G),$$

which gives rise to the isomorphism $\lambda_g : \mathbb{T}^*(G) \longrightarrow \mathbb{T}^*(G)$ with the inverse λ_g^{-1}. Define a map

$$\lambda : G \times \mathfrak{g}^* \longrightarrow \mathbb{T}^*G$$

$$(g, \xi) \longmapsto \lambda_g^{-1}(g)(\xi),$$

which is a diffeomorphism in the following commutative diagram:

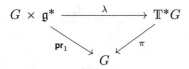

The co-tangent bundle $\mathbb{T}^*G \simeq G \times \mathfrak{g}^*$ is a trivial vector bundle with the fiber $\mathfrak{g}^*$. The Liouville form $\rho \in \Omega^1(\mathbb{T}^*G)$ defines by the section $\sigma_\rho : \mathbb{T}^*G \longrightarrow \mathbb{T}^*\mathbb{T}^*G$ similar to Sect. 1.2.4:

$$\langle \sigma_\rho(\xi), \mathcal{X} \rangle = \langle \rho, \mathbb{T}_{\pi_\xi}(\mathcal{X}) \rangle, \qquad \mathcal{X} \in \mathbb{T}_{\pi_\xi}(\mathbb{T}^*G), \qquad \xi \in \mathbb{T}_h^*G.$$

Let $g \in G, \xi \in \mathbb{T}_{gh}^*G, \mathcal{X} \in \mathbb{T}_{\pi_{\lambda_g(\xi)}}(\mathbb{T}^*G)$, then we have

$$\langle \sigma_\rho \circ \lambda_g(\xi), \mathcal{X} \rangle = \langle \lambda_g(\xi), \mathbb{T}_{\pi_{\lambda_g(\xi)}} \rangle = \langle (\mathbb{T}_h L_g)^* \xi, \mathbb{T}_{\lambda_g(\xi)} \rangle =$$

$$\langle \xi, (\mathbb{T}_h L_g) \mathbb{T}_{\lambda_g(\xi)} \mathcal{X} \rangle = \langle \xi, \mathbb{T}_{\lambda_g(\xi)} (L_g \circ \pi) \mathcal{X} \rangle.$$

We can make two observations:

- $(L_g \circ \pi)(h, \xi) = gh$,
- $\pi \circ \lambda_{g^{-1}} = gh$.

Henceforth, $L_g \circ \pi \equiv \pi \circ \lambda_{g^{-1}}$. The Liouville form becomes

$$\langle \sigma_\rho \circ \lambda_g(\xi), \mathcal{X} \rangle =$$

$$\langle \xi, \mathbb{T}_{\lambda_g(\xi)} (\pi \circ \lambda_{g^{-1}}) \mathcal{X} \rangle = \langle \xi, \mathbb{T}_\xi \pi \circ \mathbb{T}_{\lambda_g(\xi)} \lambda_{g^{-1}}(\mathcal{X}) \rangle =$$

$$\langle \rho(\xi), \mathbb{T}_{\lambda_g(\xi)} \lambda_{g^{-1}}(\xi) \mathcal{X} \rangle = \langle \lambda_{g^{-1}}^* \circ \rho(\xi), \mathcal{X} \rangle.$$

We have $\omega = d\rho$ as the canonical symplectic form on $\mathbb{T}^*G$. We observe also that

- $\sigma_\rho \circ \lambda_g = \lambda_{g^{-1}}^*(\sigma_\rho)$,
- $\omega \circ \lambda_g = \lambda_{g^{-1}}^*(\omega)$.

We recall that a Poisson bracket $\{F, G\}$ for $F, G \in C^\infty(\mathbb{T}^*G)$ can be defined via $\{F, G\} \overset{\text{def}}{:=} \mathcal{X}_F(G)$, where $\mathcal{X}_F$ is the unique vector field on $\mathbb{T}^*G$ such that $\mathcal{X}_F \lrcorner \omega = dF$.

Lemma 1.4 *Let $g \in G$ and $F, G \in C^\infty(\mathbb{T}^*G)$, then*

$$\{F \circ \lambda_g, G \circ \lambda_g\} = \{F, G\} \circ \lambda_g.$$

Proof Left to the reader as an exercise.   □

Let $C^\infty(\mathbb{T}^*G)^G$ denote a subspace of stable or invariant functions with respect to the mapping λ_g, i.e.

$$C^\infty(\mathbb{T}^*G)^G \overset{\text{def}}{:=} \{F \in C^\infty(\mathbb{T}^*G) \mid F \circ \lambda_g = F\}, \qquad g \in G.$$

Lemma 1.4 shows that the set $C^\infty(\mathbb{T}^*G)^G$ is closed with respect to the Poisson bracket.

Let the linear mapping Φ be defined as

$$\Phi : C^\infty(\mathfrak{g}^*) \longrightarrow C^\infty(\mathbb{T}^*G)$$

$$F \longmapsto F \circ \mathbf{pr}_2,$$

where $\mathbf{pr}_2 : \mathbb{T}^*G \equiv G \times \mathfrak{g}^* \longrightarrow \mathfrak{g}^*$ is the canonical projection on the second argument. Let $\iota : \mathfrak{g}^* \longrightarrow \mathbb{T}^*G$ be the canonical embedding. Then,

$$\Psi : C^\infty(\mathbb{T}^*G)^G \longrightarrow \mathfrak{g}^*$$

$$F \longmapsto F \circ \iota$$

is linear and inverse to Φ, i.e. $\Psi \equiv \Phi^{-1}$.

Lemma 1.5 *The bracket $\{F, G\} \overset{\text{def}}{:=} \Phi^{-1}(\{\Phi(F), \Phi(G)\})$ is a Poisson bracket on $C^\infty(\mathfrak{g}^*)$ coinciding with two previous definitions.*

Proof Left to the reader as an exercise.   □

1.3 Group Actions and Orbits

Let G be a Lie group and $\mathbb{M}$ is a smooth manifold.

Definition 1.4 A *left action* of G on $\mathbb{M}$ is a smooth map $\mu : G \times \mathbb{M} \longrightarrow \mathbb{M}$ such that

- $\mu(e, m) = m, \forall m \in \mathbb{M},$

- $\mu\big(g, \mu(h, m)\big) = \mu(gh, m), \forall g, h \in G$ and $\forall m \in \mathbb{M}$.

Definition 1.5 A *right action* of G on $\mathbb{M}$ is a smooth map $\rho : \mathbb{M} \times G \longrightarrow \mathbb{M}$ such that

- $\rho(m, e) = m, \forall m \in \mathbb{M}$,
- $\rho\big(\rho(m, h), g\big) = \rho(hg, m), \forall g, h \in G$ and $\forall m \in \mathbb{M}$.

Left and right actions of G and $\mathbb{M}$ are in one-to-one correspondence by the following relation:

$$\rho(m, g^{-1}) = \mu(g, m).$$

From now on we shall denote the left Lie group action $\mu(g, m)$ simply by $g \cdot m$. We can define several important action types:

Effective or Faithful $\forall g \in G, g \neq e \implies \exists m \in \mathbb{M}$ such that $g \cdot m \neq m$.
Free   If g is a group element and $\exists m \in \mathbb{M}$ such that $g \cdot m = m$ (that is, if g has at least one fixed point), $\implies g = e$. Note that a free action on a non-empty M is faithful.
Transitive   If $\forall m, n \in \mathbb{M}, \exists g \in G$ such that $g \cdot m = n$. In this case the smooth manifold $\mathbb{M}$ is called homogeneous.

Important examples of group actions include:

Example 1.1 Example G acts on itself by left multiplication:

$$G \times G \longrightarrow G$$
$$(g, h) \longmapsto gh.$$

This action is effective and transitive. Indeed, $g \cdot h = h \implies g = e$ and if $g \cdot m = n \implies g = n \cdot m^{-1}$.

Example 1.2 Example G acts on itself by conjugation:

$$G \times G \longrightarrow G$$
$$(g, h) \longmapsto g \cdot h \cdot g^{-1}.$$

Generally, this action is not free, transitive, or effective.

Example 1.3 Example $\mathbf{GL}_n(\mathbb{R})$ acts on $\mathbb{R}^n \setminus \{0\}$ by matrix multiplication on the left:

$$\mathbf{GL}_n(\mathbb{R}) \times \mathbb{R}^n \setminus \{0\} \longrightarrow \mathbb{R}^n \setminus \{0\}$$
$$(A, x) \longmapsto A x.$$

This is an example of an effective and transitive action.

1.3.1   Stabilizers and Orbits

Let G be a Lie group which acts on a smooth manifold $\mathbb{M}$. The *orbit* of a point $m \in \mathbb{M}$ is

$$G \cdot m \overset{\text{def}}{:=} \{ g \in G \mid g \cdot m \} \subseteq \mathbb{M}.$$

A *stabilizer* of a point $m \in \mathbb{M}$ is

$$G_m \overset{\text{def}}{:=} \{ g \in G \mid g \cdot m = m \} \subseteq G.$$

Proposition 1.1 *The stabilizer G_m is a closed sub-group of G and $G_{g \cdot m} = g \cdot G_m \cdot g^{-1}, \forall g \in G$.*

Proof Left to the reader as an exercise.   □

We mention here two technical theorems regarding the orbits and stabilizers:

Theorem 1.3 *Let G be a Lie group which acts on a smooth manifold $\mathbb{M}$ and $m \in \mathbb{M}$. There is a manifold structure on the orbit $G \cdot m$ such that the map*

$$G \longrightarrow G \cdot m$$

$$g \longmapsto g \cdot m$$

is a submersion and the embedding $\iota : G \cdot m \hookrightarrow \mathbb{M}$ is an immersion.

Theorem 1.4 *The Lie algebra $\mathfrak{g}_m$ of the stabilizer G_m for a point $m \in \mathbb{M}$ coincides with $\ker \mathbb{T}_e \Phi$, where the mapping Φ is defined as*

$$\Phi : G \longrightarrow \mathbb{M}$$

$$g \longmapsto g \cdot m.$$

1.3.2   Infinitesimal Action

Let $\mu : G \times \mathbb{M} \longrightarrow \mathbb{M}$ be a Lie group action on $\mathbb{M}$ and $\mathfrak{g} = \textbf{Lie}\,(G)$ be its Lie algebra.

Definition 1.6 Let $\mathcal{X} \in \mathfrak{g}$ and $\phi : \mathbb{R} \longrightarrow G$ its exponential flow, i.e. $\phi\,(t) = \exp(t\,\mathcal{X})$. Then, there exists the unique vector field $\mathcal{X}_{\mathbb{M}} \in \mathfrak{X}\,(\mathbb{M})$ with the flow $\phi_m : \mathbb{R} \longrightarrow \mathbb{M}$ defined by $\phi_m\,(t) = \phi\,(t) \cdot m$. The vector field $\mathcal{X}_{\mathbb{M}}$ is defined by

$$\mathcal{X}_{\mathbb{M}}(m)\,(f) \overset{\text{def}}{:=} \frac{\mathrm{d}}{\mathrm{d}t}\left(f \circ \phi(t) \cdot m \right)\bigg|_{t=0}.$$

The mapping $\mu_* : \mathfrak{g} \longrightarrow \mathfrak{X}(\mathbb{M})$ is called the *infinitesimal action* of $\mathfrak{g}$ on $\mathbb{M}$.

Proposition 1.2 *The mapping* $\mu_* : \mathfrak{g} \longrightarrow \mathfrak{X}(\mathbb{M})$ *is a Lie algebra (anti-)homomorphism (and it is in particular a linear mapping):*

$$\mu_*([\mathcal{X}, \mathcal{Y}]) = -[\mu_*(\mathcal{X}), \mu_*(\mathcal{Y})].$$

Proof Left to the reader as an exercise.   □

Remark 1.1 One can see that $\mu_*(\mathcal{X})_m(f) = \mathcal{X}(f \circ \Phi_m)$. In other words, $\mu_*(\mathcal{X})_m = \mathbb{T}_e \Phi_m(\mathcal{X})$.

Proposition 1.3 *Let* $m \in \mathbb{M}$*, then*

$$\mathbb{T}_m G \cdot m = \{\mathcal{X} \in \mathfrak{g} \mid \mu_*(\mathcal{X})_m\}.$$

Proof Left to the reader as an exercise.   □

The following difficult result is left without the proof:

Theorem 1.5 (R. Palais) *Let* G *be a simply connected Lie group with the Lie algebra* $\mathfrak{g} = \mathbf{Lie}(G)$ *and* $\mathbb{M}$ *be a smooth compact manifold such that there exists a homomorphism of Lie algebras* $\rho : \mathfrak{g} \longrightarrow \mathfrak{X}(\mathbb{M})$*. Then, there is a unique action* $\mu : G \times \mathbb{M} \longrightarrow \mathbb{M}$ *such that* $\mu_* = \rho$.

Proposition 1.4 *Let* $\mu : G \times \mathbb{M} \longrightarrow \mathbb{M}$ *be an action of* G *on a smooth manifold* $\mathbb{M}$ *and* $m \in \mathbb{M}$*. Then, the following diagram commutes:*

$$
\begin{array}{ccc}
\mathfrak{g} & \xrightarrow{\ \mu_*\ } & \mathfrak{X}(\mathbb{M}) \\
{\scriptstyle \exp}\big\downarrow & & \big\downarrow{\scriptstyle \exp} \\
G & \xrightarrow[\ \mu_m\ ]{} & \mathbb{M}
\end{array}
$$

or, in other words:

$$\forall \mathcal{X} \in \mathfrak{g}: \qquad \mu_m(e^{t\mathcal{X}}) = e^{t\,\mu_*(\mathcal{X})_m}.$$

1.3.3   Lie Group and Lie Algebra Representations

Let G be a Lie group, $\mathfrak{g} = \mathbf{Lie}(G)$ be its Lie algebra and V be a real vector space.

Definition 1.7 A *representation* of the Lie group G in the vector space V is a homomorphism of Lie groups (i.e. a smooth group morphism) $\varphi : G \longrightarrow \mathbf{GL}(V)$.

Definition 1.8 A representation of the Lie algebra $\mathfrak{g}$ in the vector space V is a Lie algebra homomorphism $\phi : \mathfrak{g} \longrightarrow \mathbf{End}(V)$.

Here $\mathbf{End}(V)$ is enabled with the Lie algebra structure given by the endomorphisms commutator:

$$\forall A, B \in \mathbf{End}(V) \qquad [A, B] \overset{\text{def}}{:=} A \cdot B - B \cdot A.$$

If $\varphi : G \longrightarrow \mathbf{GL}(V)$ is a Lie group representation, then

$$\phi \overset{\text{def}}{:=} \mathbb{T}_e \varphi : \mathbb{T}_e G = \mathfrak{g} \longrightarrow \mathbb{T}_{\text{id}}\big(\mathbf{GL}(V)\big) = \mathbf{End}(V)$$

is a representation of the Lie algebra $\mathfrak{g}$.

1.3.3.1 Adjoint Representations

Let $g \in G, V = \mathfrak{g}$, then the composition $L_g \circ R_{g^{-1}} : G \longrightarrow G$ induces a linear mapping $\mathbb{T}_e(L_g \circ R_{g^{-1}}) \overset{\text{def}}{:=} \mathbf{Ad}(g) : \mathfrak{g} \longrightarrow \mathfrak{g}$ and, hence, a group morphism $\mathbf{Ad} : G \longrightarrow \mathbf{GL}(\mathfrak{g})$. Then, the following Lemma holds:

Lemma 1.6 *The group morphism* $\mathbf{Ad}$ *is a smooth map which gives a representation of G in $\mathfrak{g}$, which is called the* adjoint *Lie group representation.*

Proof Left to the reader as an exercise. □

Let $\mathbf{ad} \overset{\text{def}}{:=} \mathbb{T}_e(\mathbf{Ad}) : \mathfrak{g} \longrightarrow \mathbf{End}(\mathfrak{g})$. Then, $\mathbf{ad}$ is also called the *adjoint* Lie group representation.

Lemma 1.7

$$\mathbf{ad}(\mathcal{X})(\mathcal{Y}) = [\mathcal{X}, \mathcal{Y}], \qquad \forall \mathcal{X}, \mathcal{Y} \in \mathfrak{g}.$$

Proof Left to the reader as an exercise. □

1.3.3.2 Co-Adjoint Representations

Let $g \in G, V \in \mathfrak{g}^*$ and $f^* \in \mathbf{End}(\mathfrak{g}^*)$ is defined by $f^*(\xi) \overset{\text{def}}{:=} \xi \circ f$ for any element $f \in \mathbf{End}(\mathfrak{g})$. Then, we can write down the following

Definition 1.9 The following smooth map

$$\mathbf{Ad}^* : G \longrightarrow \mathbf{GL}(\mathfrak{g}^*),$$

$$g \longmapsto \mathbf{Ad}(g^{-1})^*,$$

which gives a representation of G in $\mathfrak{g}^*$ is called the co-adjoint Lie group representation.

The last definition makes sense because $\mathbf{Ad}^* = F \circ \mathbf{Ad}$ and $F(f) = f^*$, where the map $F : \mathbf{End}(V) \longrightarrow \mathbf{End}(V^*)$. Similarly, we can also define

$$\mathbf{ad}^* : \mathfrak{g} \longrightarrow \mathbf{End}(\mathfrak{g}^*),$$
$$\mathcal{X} \longmapsto -\mathbf{ad}^*(\mathcal{X}),$$

where

$$\mathbf{ad}^*(\mathcal{X})(\xi)(\mathcal{Y}) = -\langle \xi, [\mathcal{X}, \mathcal{Y}] \rangle.$$

Then, $\mathbf{ad}^*$ is also called the co-adjoint Lie algebra representation in $\mathfrak{g}^*$.

Lemma 1.8

$$[\mathbf{ad}^*(\mathcal{X}), \mathbf{ad}^*(\mathcal{Y})] = \mathbf{ad}^*([\mathcal{X}, \mathcal{Y}]), \qquad \forall \mathcal{X}, \mathcal{Y} \in \mathfrak{g}, \quad \forall \xi \in \mathfrak{g}^*.$$

Proof Left to the reader as an exercise. □

Proposition 1.5 *The co-adjoint representation* $\mathbf{Ad}^*$ *of a Lie group G gives rise to a co-adjoint left action of G on $\mathfrak{g}^*$:*

$$G \times \mathfrak{g}^* \longrightarrow \mathfrak{g}^*,$$
$$(g, \xi) \longmapsto \mathbf{Ad}^*(\xi).$$

Let $\mathcal{X} \in \mathfrak{g}$ and $F_{\mathcal{X}} \in C^\infty(\mathfrak{g}^*)$ be the evaluation function defined by $F_{\mathcal{X}}(\xi) \overset{\text{def}}{:=} \xi(\mathcal{X})$. Then, the following Propositions hold:

Proposition 1.6

$$\frac{\mathrm{d}}{\mathrm{d}t} F\left(\mathbf{Ad}^*_{\exp(t\mathcal{X})}(\xi)\right)\bigg|_{t=0} = \{F, F_{\mathcal{X}}\}(\xi).$$

Proposition 1.7 *A function* $F \in C^\infty(\mathfrak{g}^*)$ *is a Casimir function for the Lie–Poisson structure on $\mathfrak{g}^*$ if and only if*

$$\{F, F_{\mathcal{X}}\} = 0, \qquad \forall \mathcal{X} \in \mathfrak{g}.$$

Finally, we can state without the proof the following important

Theorem 1.6 (Lie–Berezin–Kirillov–Kostant–Souriau)

$$\mathbf{Cas}\left(C^\infty(\mathfrak{g}^*)\right) = C^\infty(\mathfrak{g}^*)^{\mathbf{Ad}^*_G},$$

where

$$C^\infty(\mathfrak{g}^*)^{\mathbf{Ad}_G^*} \overset{\text{def}}{:=} \left\{ F \in C^\infty(\mathfrak{g}^*) \,\middle|\, F \circ \mathbf{Ad}_g^* = F, \quad \forall g \in G \right\}.$$

Denote by $\mathcal{O}_\xi \overset{\text{def}}{:=} G \cdot \xi \subseteq \mathfrak{g}^*$ a co-adjoint orbit of a co-vector $\xi \in \mathfrak{g}^*$. Recall that these orbits are manifolds such that $\mathcal{O}_\xi$ admits a submersion $\phi : G \longrightarrow \mathcal{O}_\xi$ and an immersion $\iota : \mathcal{O}_\xi \hookrightarrow \mathfrak{g}^*$. Define a $\mathfrak{g}$−valued 1−form ω on G as

$$\omega_g(\mathcal{X}) = \mathbb{T}_g L_{g^{-1}}(\mathcal{X}) \in \mathfrak{g}.$$

Then, $L_h^*(\omega) = \omega$. In other words, the 1−form ω is G−invariant. Here we take $g, h \in G$ and $\mathcal{X} \in \mathfrak{g}$. By Maurer–Cartan formula we have that

$$d\omega = -\frac{1}{2}[\omega, \omega].$$

Let us define also the 1−form $\omega_\xi \in \Lambda^1(G)$ by

$$\omega_\xi(g)(\mathcal{X}) \overset{\text{def}}{:=} \langle \xi, \omega_g(\mathcal{X}) \rangle.$$

Then, the following result can be shown:

Theorem 1.7 (Kirillov–Kostant–Souriau) *There exists a unique 2−form $\Omega_\xi \in \Lambda^2(\mathcal{O}_\xi)$ such that $\phi^*(\Omega_\xi) = d\omega_\xi$. This form is symplectic on the co-adjoint orbit $\mathcal{O}_\xi$.*

1.4 Moment Map, Poisson and Hamiltonian Actions

1.4.1 Introductory Motivation

Let $\mathbb{R}^3$ be a basic configuration space with coordinate or position vectors $\mathbf{r} = (q_1, q_2, q_3)$ and velocity vectors:

$$\dot{\mathbf{r}} = (\dot{q}_1, \dot{q}_2, \dot{q}_3) \overset{\text{def}}{=:} \mathbf{p} \overset{\text{def}}{:=} (p_1, p_2, p_3).$$

$$E_{\mathrm{T}} \overset{\text{def}}{:=} \frac{\langle \dot{\mathbf{r}}, \dot{\mathbf{r}} \rangle}{2} + U(\mathbf{r})$$

and the equation of motion is $\ddot{\mathbf{r}} = -\nabla_{\mathbf{r}} U(\mathbf{r})$. The total mechanical energy is conserved, i.e. $\dfrac{dE_{\mathrm{T}}}{dt} \equiv 0$. The angular momentum is also constant along a trajectory. It implies that

$$\frac{d\mathbf{L}}{dt} = \dot{\mathbf{r}} \times \dot{\mathbf{r}} + \mathbf{r} \times \ddot{\mathbf{r}} = -\mathbf{r} \times \nabla_\mathbf{r} U = \mathbf{0},$$

which is equivalent to say that $\mathbf{r} = \lambda \, \nabla_\mathbf{r} U$ for some $\lambda \in \mathbb{R}$.

Let $\mathfrak{so}\,(3)$ be the Lie algebra of skew-symmetric 3×3 matrices with real entries. This is a three-dimensional vector space with the basis $\{\, X_1, \, X_2, \, X_3 \,\}$ given by three following matrices:

$$X_1 = \begin{pmatrix} 0 & 1 & 0 \\ -1 & 0 & 0 \\ 0 & 0 & 0 \end{pmatrix}, \quad X_2 = \begin{pmatrix} 0 & 0 & -1 \\ 0 & 0 & 0 \\ 1 & 0 & 0 \end{pmatrix}, \quad X_3 = \begin{pmatrix} 0 & 0 & 0 \\ 0 & 0 & 1 \\ 0 & -1 & 0 \end{pmatrix}.$$

The Lie brackets in the Lie algebra $\mathfrak{so}\,(3)$ are given by

$$[\, X_i, \, X_j \,] = X_k, \qquad (i, \, j, \, k) = (1, \, 2, \, 3),$$

with all circular permutations. The Killing form $\kappa\,(-, \, -)$ defined as

$$\kappa : \mathfrak{so}\,(3) \times \mathfrak{so}\,(3) \longrightarrow \mathbb{R}$$

$$(X, \, Y) \longmapsto \operatorname{tr}\,(XY)$$

is symmetric, bi-linear and non-degenerate. Here $\operatorname{tr}\,(-)$ is the trace form of a square matrix. This form identifies $\mathfrak{so}\,(3)$ and $\mathfrak{so}\,(3)^*$ by the interior product rule $X \lrcorner \kappa$.

The Lie–Poisson structure on $\mathfrak{so}\,(3)^*$ in the coordinates $(x_1, \, x_2, \, x_3)$ on $\mathfrak{so}\,(3)$ can be expressed as

$$\{\, F, \, G \,\}\,(x_1, x_2, x_3) = \sum_{i, \, j, \, k = 1}^{3} c_{ij}^{k} \left(\frac{\partial F}{\partial x_i} \frac{\partial G}{\partial x_j} - \frac{\partial F}{\partial x_j} \frac{\partial G}{\partial x_i} \right) x_k.$$

Here c_{ij}^{k} is the structure constant tensor of the Lie algebra $\mathfrak{so}\,(3)$.

The angular momentum $\mathbf{L}$ s defined as a map

$$\mathbb{T}^*\mathbb{R}^3 \simeq \mathbb{R}^6 \longrightarrow \mathfrak{so}\,(3)^*$$

$$(\mathbf{q}, \mathbf{p}) \longmapsto \mathbf{q} \times \mathbf{p} = \sum_{i, \, j, \, k} (q_i \, p_j - p_i \, q_j) \, X_k.$$

The angular momentum map $\mathbf{L} : \mathbb{T}^*\mathbb{R}^3 \longrightarrow \mathfrak{so}\,(3)^*$ is a Poisson morphism.

Definition 1.10 Let $\mu : G \times \mathbb{M} \longrightarrow \mathbb{M}$ be a Lie group action on a Poisson manifold $\left(\mathbb{M}; \{\, -, \, - \,\} \right)$. This action is called a Poisson action if the map

$$\mu_g^* : C^\infty\,(\mathbb{M}, \, \mathbb{R}) \longrightarrow C^\infty\,(\mathbb{M}, \, \mathbb{R})$$

defined by $\mu_g^*(F)(m) \overset{\text{def}}{:=} F\left(\mu_g(m)\right)$ satisfies the following condition:

$$\mu_g^*\left(\{F, G\}\right)(m) = \left\{\mu_g^*(F), \mu_g^*(G)\right\}(m), \qquad \forall F, G \in C^\infty(\mathbb{M}, \mathbb{R}).$$

Let a Poisson structure $(\mathbb{M}, \pi)$ be symplectic.[1] In this case this Poisson action can be called the Hamiltonian action.

1.4.2 Momentum Map

Definition 1.11 Let $\mathfrak{g}$ be a Lie algebra and $\left(\mathbb{M}, \{-, -\}\right)$ be a Poisson manifold. A *momentum map* is a Poisson morphism $\mu : \mathbb{M} \longrightarrow \mathfrak{g}^*$. In other words, it is a smooth map μ such that for $\forall F, G \in C^\infty(\mathfrak{g}^*)$:

$$\mu^*\left(\{F, G\}_{\mathfrak{g}^*}\right) = \left\{\mu^*(F), \mu^*(G)\right\}_{\mathbb{M}}.$$

Let $\bar{\lambda} : \mathfrak{g} \longrightarrow C^\infty(\mathbb{M}, \mathbb{R})$ be a smooth linear map. Then, there is a unique map $\lambda : \mathbb{M} \longrightarrow \mathfrak{g}^*$ defined by $\bar{\lambda}$:

$$\{\lambda(m), \mathcal{X}\} = \bar{\lambda}(\mathcal{X})(m), \qquad \forall m \in \mathbb{M}, \quad \forall \mathcal{X} \in \mathfrak{g}.$$

Proposition 1.8 *Let* $\left(\mathbb{M}, \{-, -\}\right)$ *be a Poisson manifold and* $\mu : \mathbb{M} \longrightarrow \mathfrak{g}^*$ *is a smooth map. Then,* μ *is a momentum map if and only if the associated map* $\bar{\mu} : \mathfrak{g} \longrightarrow C^\infty(\mathbb{M}, \mathbb{R})$ *is a Lie algebra homomorphism:*

$$\bar{\mu}\left([\mathcal{X}, \mathcal{Y}]\right) = \{\bar{\mu}(\mathcal{X}), \bar{\mu}(\mathcal{Y})\}_{\mathbb{M}}, \qquad \forall \mathcal{X}, \mathcal{Y} \in \mathfrak{g}.$$

Recall that the map $\chi : C^\infty(\mathbb{M}, \mathbb{R}) \longrightarrow \mathfrak{X}(\mathbb{M})$ such that

$$\chi(F) = \mathcal{X}_F = \{F, -\}$$

is a Lie algebra morphism. We take the composition $\Theta \overset{\text{def}}{:=} \chi \circ \bar{\mu} : \mathfrak{g} \longrightarrow \mathfrak{X}(\mathbb{M})$, where $\bar{\mu} : \mathfrak{g} \longrightarrow C^\infty(\mathbb{M}, \mathbb{R})$ and $\mathfrak{g} = \mathbf{Lie}(G)$ with a simply connected Lie group G. For compact manifolds $\mathbb{M}$, Theorem 1.5 ensures the existence of an action $\lambda : G \times \mathbb{M} \longrightarrow \mathbb{M}$ with $\lambda_* = -\Theta$.

Proposition 1.9 *If* G *is connected, then* $\lambda_* = -\Theta$ *gives a Poisson morphism* $\lambda_g^* : C^\infty(\mathbb{M}, \mathbb{R}) \longrightarrow C^\infty(\mathbb{M}, \mathbb{R})$ *for* $\forall g \in G$ *and* $\forall u, v \in C^\infty(\mathbb{M}, \mathbb{R})$:

---

[1] Here we mean that the bi-vector π is non-degenerate, i.e. π is invertible when it is seen as a banal matrix.

$$\left\{ \lambda_g^*(u), \lambda_g^*(v) \right\} = \lambda_g^*(\{u, v\}).$$

Proposition 1.10 *Let $\mathbb{M}$ be a compact manifold and G is connected and simply connected. Then, the action λ is $G-$equivariant:*

$$
\begin{array}{ccc}
\mathbb{M} & \xrightarrow{\ \mu\ } & \mathfrak{g}^* \\
\lambda_g \downarrow & & \downarrow Ad_g^* \\
\mathbb{M} & \xrightarrow{\ \mu\ } & \mathfrak{g}^*
\end{array}
$$

1.4.3 Moment and Hamiltonian Actions

Let $(\mathbb{M}, \omega)$ be a symplectic manifold and the corresponding Poisson brackets are defined by a pair of Hamiltonian vector fields:

$$\{u, v\} = \mathcal{X}_u(v) = \omega(\mathcal{X}_v, \mathcal{X}_u), \qquad \mathcal{X}_u \lrcorner \omega = du.$$

Lemma 1.9 *If $H^1(\mathbb{M}, \mathbb{R}) = 0$ and $\mathcal{X} \in \mathfrak{X}(\mathbb{M})$ "infinitesimally" preserves the symplectic form ω, i.e. $\mathcal{L}_{\mathcal{X}}(\omega) = 0$, then there exists a unique $u \in C^\infty(\mathbb{M}, \mathbb{R})$ such that $\mathcal{X} = \mathcal{X}_u$.*

Here we should remark that the function u is uniquely defined only modulo a locally constant function on M (which is usually identified with an element of $H^0(\mathbb{M}, \mathbb{R})$).

Lemma 1.10 *Let $\lambda : G \times \mathbb{M} \longrightarrow \mathbb{M}$ be an action of a Lie group G on a symplectic manifold $(\mathbb{M}, \omega)$. The action λ is a Poisson (more precisely, in this case we may call it a Hamiltonian) action if and only if $\lambda_g^* = \omega$.*

Proposition 1.11 *Let $\lambda : G \times \mathbb{M} \longrightarrow \mathbb{M}$ be a Hamiltonian action on a symplectic manifold $(\mathbb{M}, \omega)$ and $\lambda_* : \mathfrak{g} \longrightarrow \mathfrak{X}(M)$ is the corresponding Lie algebra homomorphism. Then, $\forall \mathcal{X} \in \mathfrak{g}$:*

$$\mathcal{L}_{\lambda_*(\mathcal{X})}(\omega) = 0.$$

Definition 1.12 Let $\lambda : G \times \mathbb{M} \longrightarrow \mathbb{M}$ be an action of G on $\mathbb{M}$ and $(\mathbb{T}^*\mathbb{M}, \Omega)$ is the co-tangent bundle with the canonical symplectic form $\Omega = d\rho$, where ρ is the Liouville $1-$form. This action can be lifted to an action $\Lambda : G \times \mathbb{T}^*\mathbb{M} \longrightarrow \mathbb{T}^*\mathbb{M}$ defined by

$$\Lambda(g, \xi_m) \overset{\text{def}}{:=} (\mathbb{T}_{g \cdot m} \lambda_{g^{-1}}^*)(\xi_m).$$

Theorem 1.8 *The action Λ is Hamiltonian and the induced momentum map μ_Λ : $\mathbb{T}^*\mathbb{M} \longrightarrow \mathfrak{g}^*$ is defined by*

$$\{\mu_\Lambda(\xi_m), \mathcal{X}\} = \{\xi_m, \mathbb{T}_e\lambda_m(\mathcal{X})\}.$$

1.4.3.1 Examples

Example 1.4 Lifting of the left $G-$action *on* G to $\mathbb{T}^*G$:

$$\lambda : G \times G \longrightarrow G,$$
$$(g, h) \longmapsto g \cdot h.$$

Then, we obtain the required lifting:

$$\Lambda : G \times \mathbb{T}^*G \simeq G \times \mathfrak{g}^* \longrightarrow \mathbb{T}^*G \simeq G \times \mathfrak{g}^*,$$
$$\big(g, (h, \xi)\big) \longmapsto (g \cdot h, \xi).$$

The associated momentum can be also easily computed:

$$\mu(\xi_h) = -(\mathbb{T}_e R_h)^*(\xi_h), \qquad \mu(h, \xi) = \mathbf{Ad}^*_h(\xi).$$

Similarly, we can consider lifting of the right $G-$action *on* G to $\mathbb{T}^*G$:

$$\lambda : G \times G \longrightarrow G,$$
$$(g, h) \longmapsto h \cdot g^{-1}.$$

Then, we obtain the required lifting:

$$\Lambda : G \times \mathbb{T}^*G \simeq G \times \mathfrak{g}^* \longrightarrow \mathbb{T}^*G \simeq G \times \mathfrak{g}^*,$$
$$\big(g, (h, \xi)\big) \longmapsto (h \cdot g^{-1}, \mathbf{Ad}^*_g\xi).$$

The associated momentum can be also easily computed:

$$\mu(\xi_h) = -(\mathbb{T}_e L_h)^*(\xi_h), \qquad \mu(h, \xi) = -\xi.$$

Example 1.5 Let us consider also the action of $\mathbb{S}^1$ on $\mathbb{C}$:

$$\mathbb{C} \simeq \mathbb{R}^2 \simeq \mathbb{T}^*\mathbb{R}^1, \qquad \Omega = \mathrm{d}q \wedge \mathrm{d}p.$$

The action is given by

$$\lambda : \mathbb{S}^1 \times \mathbb{C} \longrightarrow \mathbb{C},$$
$$\left(e^{i\theta}, z \right) \longmapsto e^{i\theta} z,$$

for some $\theta \in [0, 2\pi[$. Above i is the complex imaginary unit, i.e. $i^2 = -1$. Then, one can easily obtain the expression for $\lambda_* : \mathbb{T}_e \mathbb{S}^1 \longrightarrow \mathbb{C}$:

$$\lambda_* \left(\frac{\mathrm{d}}{\mathrm{d}\theta} \right) (q, p) = \mathbb{T}_e \left(\lambda_{q, p} \right) \left(\frac{\mathrm{d}}{\mathrm{d}\theta} \right) = -p \frac{\partial}{\partial q} + q \frac{\partial}{\partial p} .$$

The interior product with the symplectic form can be also easily obtained:

$$\lambda_* \left(\frac{\mathrm{d}}{\mathrm{d}\theta} \right) \lrcorner \Omega = -(p\mathrm{d}p + q\mathrm{d}q) = -\frac{1}{2} \mathrm{d}(p^2 + q^2).$$

The momentum map is given by

$$\mu(z) = \mu(q + ip) = \frac{p^2 + q^2}{2}.$$

The last construction can be easily generalized to $\mathbb{C}^n$:

$$\lambda : \mathbb{S}^1 \times \mathbb{C}^n \longrightarrow \mathbb{C}^n,$$
$$\left(e^{i\theta}, (z_1, z_2, \ldots, z_n) \right) \longmapsto \left(e^{i\theta} z_1, e^{i\theta} z_2, \ldots, e^{i\theta} z_n \right).$$

Then, the associated momentum map is given by

$$\mu(z_1, z_2, \ldots, z_n) = \sum_{i=1}^{n} |z_i|^2.$$

Example 1.6 Let us consider the action of $\mathbb{S}^1$ on $\mathbb{S}^2$. The manifold $\mathbb{S}^2$ is equipped with local coordinates (z, ϕ) and $\Omega = \mathrm{d}z \wedge \mathrm{d}\phi$. The action of $\mathbb{S}^1$ is given by rotation in $z-$planes:

$$\lambda : \mathbb{S}^1 \times \mathbb{S}^2 \longrightarrow \mathbb{S}^2,$$
$$\left(e^{i\theta}, (z, \phi) \right) \longmapsto (z, \phi + \theta).$$

It is not difficult to see that

$$\lambda_* \left(\frac{\mathrm{d}}{\mathrm{d}\theta} \right) (z, \phi) = \frac{\partial}{\partial \phi}, \qquad \lambda_* \left(\frac{\mathrm{d}}{\mathrm{d}\theta} \right) \lrcorner \Omega = \mathrm{d}z.$$

Finally, the momentum map is

$$\mu(z, \phi) = z.$$

Example 1.7 We consider now the action of $\mathbb{S}^1$ on the torus $\mathbb{T}^2 \overset{\text{def}}{:=} \mathbb{S}^1 \times \mathbb{S}^1$. The torus $\mathbb{T}^2$ is equipped with local coordinates (ϕ_1, ϕ_2) and the symplectic form is $\Omega = d\phi_1 \wedge d\phi_2$. The action is defined as

$$\lambda : \mathbb{S}^1 \times \mathbb{T}^2 \longrightarrow \mathbb{T}^2,$$

$$\left(e^{i\theta}, \left(e^{i\phi_1}, e^{i\phi_2}\right)\right) \longmapsto \left(e^{i\phi_1}, e^{i(\phi_1 + \theta)}\right).$$

Then, we have

$$\lambda_* \left(\frac{d}{d\theta}\right)(\Omega) = d\phi_1 \wedge d(\theta + \phi_2) = \Omega$$

and

$$\lambda_* \left(\frac{d}{d\theta}\right) \lrcorner \Omega = -d\phi_1.$$

Since the coordinate function ϕ_1 is defined only locally, the momentum map μ and the morphism $\bar{\mu}$ do not exist.

Example 1.8 In this example we consider the action of $\mathbf{SU}(n)$ on $\mathbb{T}^*(\mathfrak{su}(n))$. We remind that $\mathbf{SU}(n)$ is the Lie group of special unitary matrices with complex coefficients:

$$\mathbf{SU}(n) \overset{\text{def}}{:=} \left\{ \mathbb{A} \in \text{Mat}_n(\mathbb{C}) \,\big|\, \mathbb{A}\mathbb{A}^* = \mathbb{I}, \, \det(\mathbb{A}) = 1 \right\},$$

where $\mathbb{I}$ is the identity matrix and $\mathbb{A}^*$ is the conjugate (or Hermitian) transpose of $\mathbb{A}$. The corresponding Lie algebra is defined as

$$\mathbf{Lie}\left(\mathbf{SU}(n)\right) = \mathfrak{su}(n) \overset{\text{def}}{:=} \{\mathbb{A} \in \text{Mat}_n(\mathbb{C}) \,|\, \mathbb{A}^* = -\mathbb{A}, \, \text{tr}(\mathbb{A}) = 0\}.$$

The Lie algebra $\mathfrak{su}(n)$ is an example of a semi-simple Lie algebra with a Killing form $\kappa(X, Y) = 2n \, \text{tr}(XY)$ and $\mathfrak{su}(n) \simeq \mathfrak{su}(n)^*$. The action is defined as

$$\lambda : \mathbf{SU}(n) \times \mathbb{T}^*(\mathfrak{su}(n)) \longrightarrow \mathbb{T}^*(\mathfrak{su}(n)),$$

$$(g, (X, L)) \longmapsto (g X g^{-1}, g L g^{-1}).$$

Then, $\Omega = \text{tr}(dX \wedge dL)$ and the momentum map is given by

$$\mu(X, L) = [X, L].$$

1.5 Reduction of the Phase Space

Let $(\mathbb{M}, \omega)$ be a symplectic manifold and $\lambda : G \times \mathbb{M} \longrightarrow \mathbb{M}$ is a Hamiltonian action, i.e.

$$\lambda_g^*(\omega) = \omega, \qquad \forall g \in G.$$

We justify the terminology by the following observation:

Lemma 1.11 *Assume that there exists a momentum map* $\mu : \mathbb{M} \longrightarrow \mathfrak{g}^*$, *one necessarily obtains that*

$$\lambda^*(\mathcal{Y}) = -\mathcal{X}_{\bar{\mu}(\mathcal{Y})}, \qquad \forall \mathcal{Y} \in \mathfrak{g}.$$

Definition 1.13 An element $c \in \mathfrak{g}^*$ is called a *regular* if $\mathbb{M}_c \overset{\text{def}}{:=} \mu^{-1}(c)$ is a sub-manifold in $\mathbb{M}$ and if

$$\ker(\mathbb{T}_m \mu) = \mathbb{T}_m \mathbb{M}_c, \qquad \forall c \in \mathbb{M}_c.$$

Lemma 1.12 *Let* $G_c \overset{\text{def}}{:=} \left\{ g \in G \,\middle|\, \boldsymbol{Ad}_g^*(c) = c \right\}$. *If* G *is connected and simply connected, then* $\forall m \in \mathbb{M}_c$ *and* $\forall g \in G_c$:

$$g \cdot m \in \mathbb{M}_c.$$

Proof Left to the reader as an exercise. $\qquad\qquad\qquad\qquad\qquad\qquad\qquad\square$

Remark 1.2 In the case when c is a regular element and G is connected and simply connected, the action of G on $\mathbb{M}$ induces an action of the Lie sub-group $G_c \subseteq G$ on the sub-manifold $\mathbb{M}_c \subseteq \mathbb{M}$.

1.5.1 The Main Results

Theorem 1.9 *If G is connected and simply connected and, in addition:*

- *c is a regular element;*
- *G_c is compact;*
- *G_c acts on $\mathbb{M}_c$ by free and transitive action.*

Then, there exists a natural smooth structure on $\mathbb{M}_c / G_c$ such that the mapping $\pi_c : \mathbb{M}_c \longrightarrow \mathbb{M}_c / G_c$ *is a submersion.*

Remark 1.3 The quotient space $\mathbb{M}_c / G_c$ is called in this case the *reduced phase space*.

Theorem 1.10 (Marsden–Weinstein) *If G is connected and simply connected and, in addition:*

- *c is a regular element;*
- *G_c is compact;*
- *G_c acts on $\mathbb{M}_c$ by free and transitive action.*

Then, there exists a unique symplectic 2–form ω_c on $\mathbb{M}_c / G_c$ such that

$$\pi_c^* (\omega_c) = \iota_c^* (\omega),$$

where $\pi_c : \mathbb{M}_c \longrightarrow \mathbb{M}_c / G_c$ is the canonical submersion and $\iota_c : \mathbb{M}_c \hookrightarrow \mathbb{M}$ is the canonical embedding.

The proof of this Theorem is based on the following

Lemma 1.13 *Let $m \in \mathbb{M}_c$. Then, $\mathbb{T}_m \mathbb{M}_c = (\mathbb{T}_m G \cdot m)^{\perp}$. In other words,*

$$\mathbb{T}_m \mathbb{M}_c = \{ \mathcal{X} \in \mathbb{T}_m \mathbb{M} \mid \omega_m (\mathcal{X}, \mathcal{Y}) = 0, \forall \mathcal{Y} \in \mathbb{T}_m G \cdot m \}.$$

Proof Left to the reader as an exercise.   □

Remark 1.4 Observe that $\mathbb{T}_m \mathbb{M}_c \cap \mathbb{T}_m G \cdot m \neq \varnothing$. More precisely, $\mathbb{T}_m \mathbb{M}_c \cap \mathbb{T}_m G \cdot m = \mathbb{T}_m G_c \cdot m$.

Corollary 1.1 *Let $\mathcal{X}_1, \mathcal{X}_2, \mathcal{Y}_1, \mathcal{Y}_2 \in \mathbb{T}_m \mathbb{M}_c \subseteq \mathbb{T}_m \mathbb{M}$ such that*

$$\mathbb{T}_m \pi_c (\mathcal{X}_1) = \mathbb{T}_m \pi_c (\mathcal{X}_2),$$
$$\mathbb{T}_m \pi_c (\mathcal{Y}_1) = \mathbb{T}_m \pi_c (\mathcal{Y}_2),$$

then, $\omega_m (\mathcal{X}_1, \mathcal{Y}_1) = \omega_m (\mathcal{X}_2, \mathcal{Y}_2)$.

Lemma 1.14 *Let $m, n \in \mathbb{M}_c$ such that $\pi_c (m) = \pi_c (n)$ and $\mathcal{X}_1, \mathcal{X}_2, \mathcal{Y}_1, \mathcal{Y}_2 \in \mathbb{T}_m \mathbb{M}_c \subseteq \mathbb{T}_n \mathbb{M}_c \subseteq \mathbb{T}_n \mathbb{M}$ such that*

$$\mathbb{T}_m \pi_c (\mathcal{X}_1) = \mathbb{T}_n \pi_c (\mathcal{X}_2),$$
$$\mathbb{T}_m \pi_c (\mathcal{Y}_1) = \mathbb{T}_n \pi_c (\mathcal{Y}_2),$$

then, $\omega_m (\mathcal{X}_1, \mathcal{Y}_1) = \omega_n (\mathcal{X}_2, \mathcal{Y}_2)$.

Proof Left to the reader as an exercise.   □

Lemma 1.15 *Let c be a regular element and $\mathcal{O}_c \overset{\text{def}}{:=} G \cdot c$ be its co-adjoint orbit. Then, $\mu^{-1} (\mathcal{O}_c)$ is a sub-manifold in $\mathbb{M}$.*

Proof Left to the reader as an exercise.   □

Theorem 1.11 *The mapping*

$$\phi : \mathcal{O}_c \longrightarrow \mathbb{M}_c / G_c ,$$

$$m \longmapsto \pi_c(g^{-1}m),$$

where $\mu(m) = \mathbf{Ad}^*_g(c)$ *is correctly defined, induces a diffeomorphism:*

$$\Phi : \pi\left(\mu^{-1}(\mathcal{O}_c)\right) \longrightarrow \mathbb{M}_c / G_c .$$

1.5.2   Example

In this Section we consider again the action of $\mathbb{S}^1$ on $\mathbb{C}^n$, which is defined as

$$\lambda : \mathbb{S}^1 \times \mathbb{C}^n \longrightarrow \mathbb{C}^n ,$$

$$\left(e^{i\theta}, \mathbf{q} + i\mathbf{p}\right) \longmapsto e^{i\theta}\mathbf{q} + ie^{i\theta}\mathbf{p},$$

where $\mathbf{p}, \mathbf{q} \in \mathbb{R}^n$. The momentum map is

$$\mu : \mathbb{C}^n \longrightarrow \mathbf{Lie}(\mathbb{S}^1) ,$$

$$\mathbf{q} + i\mathbf{p} \longmapsto -\sum_{i=1}^{n} \frac{q_i^2 + p_i^2}{2} .$$

Then, $\mathbb{M}_c = \left\{ z \in \mathbb{C}^n \mid \sum_{i=1}^n |z_i|^2 = 2c \right\} \simeq \mathbb{S}^n , c > 0$. It is also clear that $G_c \simeq \mathbb{S}^1$ and $\mathbb{S}^1$ is an Abelian group. Thus, we have

$$L_g R_{g^{-1}} = \mathbb{I} \quad \Longrightarrow \quad \mathbf{Ad}_g = \mathbf{Ad}^*_g = \mathbb{I}.$$

Henceforth,

$$\mathbb{M}_c / G_c = \mathbb{S}^n / \mathbb{S}^1 \simeq \mathbb{P}^{n-1}.$$

1.6   Poisson–Lie Groups

A Lie group G is called *Poisson–Lie group* if it is a Poisson manifold such that the multiplication $m : G \times G \longrightarrow G$ is a morphism of Poisson manifolds. Let $\mathfrak{g}$ be Lie algebra, $\mathfrak{g}^*$ be dual vector space to $\mathfrak{g}$.

Definition 1.14 We say that $\mathfrak{g}$ is a *Lie bi-algebra* if there is a Lie algebra structure $[-, -]_*$ on $\mathfrak{g}^*$ such that the map $\delta : \mathfrak{g} \longrightarrow \Lambda^2 \mathfrak{g}$ (called the *co-bracket*), dual to the bracket $[-, -]_* : \Lambda^2 \mathfrak{g}^* \longrightarrow \mathfrak{g}^*$ is a 1−cocycle with respect to the adjoint action of $\mathfrak{g}$ on $\Lambda^2 \mathfrak{g}$.

1.6.1 Modified Classical Yang–Baxter Equation

Let G be connected and a simply connected Lie group, and let $\mathfrak{g}$ be its Lie algebra. Then there is one-to-one correspondence between Poisson–Lie group structures on G and Lie bi-algebra structures on $\mathfrak{g}$.

As V. Drinfel'd showed [3], every structure on a semi-simple connected G has the following form:

$$\pi(g) = \Lambda^2\left((\mathcal{L}_g)_*\right)(\mathbf{r}) - \Lambda^2\left((\mathcal{R}_g)_*\right)(\mathbf{r}), \tag{1.2}$$

where $(\mathcal{L}_g)_*$ and $(\mathcal{R}_g)_*$ denote tangent maps of left and right translations by $g \in G$. The element $\mathbf{r} \in \Lambda^2 \mathfrak{g}$ satisfies the following condition:

$$[\![\mathbf{r}, \mathbf{r}]\!] \overset{\text{def}}{:=} [\mathbf{r}_{12}, \mathbf{r}_{13}] + [\mathbf{r}_{12}, \mathbf{r}_{23}] + [\mathbf{r}_{13}, \mathbf{r}_{23}] \in \Lambda^3 \mathfrak{g}, \tag{1.3}$$

where the right-hand side is invariant under the adjoint action of $\mathfrak{g}$. The condition (1.3) is called a *modified Yang–Baxter equation* and the bracket

$$[\![-, -]\!] : \Lambda^2 \mathfrak{g} \otimes \Lambda^2 \mathfrak{g} \longrightarrow \Lambda^3 \mathfrak{g}$$

is a so-called *Schouten–Nijenhuis bracket*. This is the natural graded (or *super-*) Lie algebra structure on the exterior algebra

$$\Lambda^{\bullet} \mathfrak{g} = \bigoplus_k \Lambda^k \mathfrak{g}.$$

Here $\mathbf{r}_{12}$, to give an example, denotes an element $\mathbf{r}_{12} = \mathbf{r} \otimes \mathbb{I}_3 \in (\mathfrak{g} \otimes \mathbf{k})^{\otimes 3}$; $\mathbf{k} \in \{\mathbb{R}, \mathbb{C}\}$ and $\mathbf{r}$ being usually called a *classical* $\mathbf{r}$*−matrix*.

The condition (1.3) ensures that the bracket $\{-, -\}_*$ on $\mathfrak{g}^*$ satisfies the Jacobi identity. The corresponding Lie bi-algebra structure is calculated in the obvious way. Namely, the co-bracket δ is given by

$$\delta(x) = \mathrm{d}_e \pi(x) = \mathcal{L}_{\bar{x}} \pi(e) = \frac{\mathrm{d}}{\mathrm{d}t} \mathbf{r}_{(\mathrm{e}^{-tx})_*} \pi(\mathrm{e}^{tx})\Big|_{t=0} = \mathbf{ad}_x(\mathbf{r}),$$

where $\mathrm{d}_e \pi$ is the intrinsic derivative of a poly-vector field on G with $\pi(e) = 0$, $\bar{x}$ is any vector field on G with $\bar{x}(e) = x$, and $\mathcal{L}_{\bar{x}}$ denotes the Lie derivative [8].

The Poisson structures of the form (1.2) are called *co-boundary* or **r**−*matrix structures*. Since for a connected semi-simple or a compact Lie group G every 1−cocycle is a co-boundary, one has the following

Proposition 1.12 . *The Poisson–Lie structures on a connected semi-simple or a compact Lie group G are in one-to-one correspondence with the solutions* $\mathbf{r} \in \Lambda^2 \mathfrak{g}$ *of the modified Yang–Baxter equation.*

1.6.2 Manin Triples

Let $\mathfrak{g}$ be a Lie bi-algebra. There is a unique Lie algebra structure on the vector space $\mathfrak{g} \oplus \mathfrak{g}^*$ such that

1. $\mathfrak{g}$ and $\mathfrak{g}^*$ are Lie sub-algebras.
2. The symmetric bi-linear form on $\mathfrak{g} \oplus \mathfrak{g}^*$ given by the relation

$$\langle \mathcal{X} + \xi, \mathcal{Y} + \eta \mathcal{Y} \rangle = \langle \mathcal{X}, \eta \rangle + \langle \mathcal{Y}, \xi \rangle, \quad \forall \mathcal{X}, \mathcal{Y} \in \mathfrak{g}, \quad \forall \xi, \eta \in \mathfrak{g}^*$$

is invariant.

This structure is given by

$$\{\mathcal{X}, \xi\} = -\mathbf{ad}^*_{\mathcal{X}}(\xi) + \mathbf{ad}^*_{\xi}(\mathcal{X}),$$

for $\mathcal{X} \in \mathfrak{g}$ and $\xi \in \mathfrak{g}^*$, where $\mathbf{ad}^*$ is the co-adjoint action. This Lie algebra is denoted by $\mathfrak{g} \bowtie \mathfrak{g}^*$ and $(\mathfrak{g} \bowtie \mathfrak{g}^*, \mathfrak{g}, \mathfrak{g}^*)$ is an example of a Manin triple. In general, a Manin triple is a decomposition of a Lie algebra $\mathfrak{g}$ with a non-degenerate invariant scalar product $\langle\ ,\ \rangle$ into direct sum of isotropic with respect to $\langle\ ,\ \rangle$ vector spaces, $\mathfrak{g} = \mathfrak{g}_+ \oplus \mathfrak{g}_-$ such that $\mathfrak{g}_\pm$ are Lie sub-algebras of $\mathfrak{g}$. It is well-known that there is one-to-one correspondence between Lie bi-algebras and Manin triples. These triples were introduced by V. Drinfel'd [4] and named after Yu. I. Manin.

1.6.3 Poisson–Lie Duality

Let G be a connected and simply connected Poisson–Lie group, $\mathfrak{y} = \mathbf{Lie}(G)$ its Lie algebra and $(\mathfrak{g} \bowtie \mathfrak{g}^*, \mathfrak{g}, \mathfrak{g}^*)$ the Manin triple. By duality, $(\mathfrak{g}^* \bowtie \mathfrak{g}, \mathfrak{g}^*, \mathfrak{g})$ is also a Manin triple. Then $\mathfrak{g}^*$ is a Lie bi-algebra. This enables us to consider a connected and simply connected Lie group G^* with a Poisson–Lie structure π^* and with the tangent Lie bi-algebra $\mathfrak{g}^*$. The Poisson–Lie group (G^*, π^*) is called the *Poisson–Lie dual* to (G, π).

1.6.4 Example of Non-Hamiltonian Action

Let G be a Poisson–Lie group with a multiplicative Poisson tensor π_g and $\mathbb{M}$ be a smooth Poisson manifold with a Poisson structure given by $\pi_{\mathbb{M}}$. Then, the product $G \times \mathbb{M}$ can be considered as a Poisson manifold with the direct sum structure $\tilde{\pi}$.

Proposition 1.13 *An action* $\sigma : G \times \mathbb{M} \longrightarrow \mathbb{M}$ *of a Poisson–Lie group G on a Poisson manifold $\mathbb{M}$ is a Poisson–Lie action if and only if*

$$\pi_{\mathbb{M}}(g \cdot m) = \Lambda^2\big((\sigma_g)_*\big)\big(\pi_{\mathbb{M}}(m)\big) + \Lambda^2\big((\sigma_m)_*\big)\big(\pi_G(g)\big).$$

Remark 1.5 One can consider any Lie group G as a Poisson–Lie group with $\pi_G \equiv 0$ then the action σ is a Poisson (action) if it gives a Poisson morphism $\pi_{\mathbb{M}}(g \cdot m) = (\sigma_g)_*\big(\pi_{\mathbb{M}}(m)\big)$.

Definition 1.15 The action $\sigma : G \times \mathbb{M} \longrightarrow \mathbb{M}$ is called a Poisson–Lie action if $\pi^* : C^\infty(\mathbb{M}) \longrightarrow C^\infty(G \times \mathbb{M})$ is a Poisson morphism:

$$\pi^*\big(\{F, H\}_{\pi_{\mathbb{M}}}\big) = \big\{\pi^*(F), \pi^*(H)\big\}_{\tilde{\pi}}.$$

Infinitesimally, a Poisson–Lie action of a Lie bi-algebra $\mathfrak{g}$ on a Poisson manifold $(\mathbb{M}, \{-, -\})$ is given by an action

$$\rho : \mathfrak{g} \longrightarrow \mathfrak{X}(\mathbb{M}),$$
$$\mathcal{X} \longmapsto V_{\mathcal{X}},$$

with $\mathcal{X} \in \mathfrak{g}$ such that

$$V_{\mathcal{X}}\{f, g\}(m) = \{V_{\mathcal{X}}f, g\}(m) + \{f, V_{\mathcal{X}}g\}(m)$$
$$- \big\{\mathcal{X}, \big[\rho^* df(m), \rho^* dg(m)\big]\big\},$$

where $\rho^* df(m) \in \mathfrak{g}^*$ and $\langle \mathcal{X}, \rho^* df(m) \rangle = V_{\mathcal{X}}f(m)$. In other words,

$$\langle \mathcal{X}, \big[\tilde{\rho}(dF)(m), \tilde{\rho}(dG)(m)\big]_*\rangle = \langle dF, dG \rangle\big(\rho(\mathcal{X})\big)(m)$$

define a *Lie algebroid* structure on $\mathbb{T}^*\mathbb{M}$.

Example 1.9 There are natural left and right actions of dual Poisson–Lie group G^* on G. These actions are called *left (right) dressing* transformations. The dressing transformations are not Hamiltonian as Semenov-Tian-Shansky proved but these actions are genuine Poisson–Lie actions [9].

Acknowledgments D.D. was supported by the Laboratory of Mathematics (LAMA UMR #5127) and the University Savoie Mont Blanc to attend the meeting in Wisła. The work of D.D. has been also supported by the French National Research Agency, through Investments for Future Program (ref. ANR−18−EURE−0016—Solar Academy). V.R. acknowledges a partial support of the project IPaDEGAN (H2020-MSCA-RISE-2017), Grant Number 778010, and of the Russian Foundation for Basic Research under the Grants RFBR 18−01−00461 and 16−51−53034−716 GFEN. Both authors would like to thank the Baltic Mathematical Institute for organizing this scientific event and the anonymous Referee who helped us to improve the presentation indicating some shortcomings and misprints.

References

1. Abraham, R., Marsden, J.E.: Foundations of Mechanics, 2nd ed. Addison-Wesley Pub. Co., Redwood City, CA (1987)
2. Arnold, V.I.: Mathematical Methods of Classical Mechanics, 2nd ed. Springer, New York (1997)
3. Drinfel'd, V.G.: Hamiltonian structures on Lie groups, Lie algebras and the geometric meaning of the classical Yang-Baxter equations. Dokl. Akad. Nauk SSSR **268**, 285–287 (1983)
4. Drinfel'd, V.G.: Quantum groups. J. Math. Sci. **41**, 898–915 (1988)
5. Fernandes, R.L.: Poisson vs. Symplectic Geometry. CIM Bulletin **20**, 15–19 (2006)
6. Kosmann-Schwarzbach, Y.: Lie bialgebras, Poisson Lie groups and dressing transformations. In: Kosmann-Schwarzbach, Y., Grammaticos, B., Tamizhmani, K.M. (eds.) Integrability of Nonlinear Systems, pp. 104–170. Springer, Berlin, Heidelberg (1997)
7. Lichnerowicz, A.: Les variétés de Poisson et leurs algèbres de Lie associées. J. Diff. Geom. **12**, 253–300 (1977)
8. Lu, J.-H., Weinstein, A.: Poisson Lie groups, dressing transformations, and Bruhat decompositions. J. Diff. Geom. **31**, 501–526 (1990)
9. Semenov-Tian-Shansky, M.A.: Dressing Transformations and Poisson Group Actions. Publ. RIMS, Kyoto Univ. **21**, 1237–1260 (1985)
10. Souriau, J. M.: Structure of Dynamical Systems: a Symplectic View of Physics. Birkäuser, Boston, MA (1997)
11. Tortorella, A.G.: A course on symplectic and contact geometry: an introduction to the Hamiltonian formalism. Tech. rep., Diffiety Institute, Lizzano-in-Belvedere (2019) https://drive.google.com/file/d/1A79IhEUq1ESDcUODl4-USti58OdH61HW
12. Vinogradov, A.M., Kupershmidt, B.A.: The structures of Hamiltonian mechanics. Rus. Math. Surv. **32**, 177–243 (1977)
13. Weinstein, A.: The local structure of Poisson manifolds. J. Diff. Geom. **18**, 523–557 (1983)
14. Weinstein, A.: Poisson geometry. Differential Geometry and its Applications **9**, 213–238 (1998)

Chapter 2
Notes on Tractor Calculi

Jan Slovák and Radek Suchánek

This is a survey article based on the lectures given by the first author at the Summer School Wisła 19, 19–29 August 2019, captured by the second author. The exposition aims at quick understanding of basic principles, omitting many proofs or at least their details. The reader might find a lot of further information in the cited sources throughout the text. In particular, our approach has been heavily inspired by Curry and Gover [12], while the general background on Cartan geometries including the tractors can be found in [11].

The six sections of the article roughly correspond to the six lectures (about 100 min each). We first introduce some elements of the tractor calculus in quite general situation. Then we focus rather on the overall structure of the invariant linear differential operators and we do not present much of the tractor calculus itself. In this sense, these lecture notes are complementary to [12], where the reader should look for the genuine calculus.

The audience was assumed to have basic knowledge of differential geometry as well as some representation theory (Lie groups and algebras, their representations, principal and associated bundles, connections, tensors, etc.). All this background can be found, e.g., in [17] and [11].

J. Slovák (✉)
Department of Mathematics and Statistics, Faculty of Science, Masaryk University, Brno, Czech Republic
e-mail: slovak@muni.cz

R. Suchánek
Department of Mathematics and Statistics, Faculty of Science, Masaryk University, Brno, Czech Republic

LAREMA UMR 6093, CNRS and Université d'Angers, Angers, France

© The Author(s), under exclusive license to Springer Nature Switzerland AG 2021
M. Ulan, E. Schneider (eds.), *Differential Geometry, Differential Equations, and Mathematical Physics*, Tutorials, Schools, and Workshops in the Mathematical Sciences, https://doi.org/10.1007/978-3-030-63253-3_2

2.1 Tracy Thomas' Conformal Tractors

Let us start with a quick review of the two very well known geometries, the Riemannian and the conformal Riemannian ones.

2.1.1 Riemannian Sphere

There are many ways how to view the standard sphere

$$S^n = \{x \in \mathbb{R}^n \mid \|x\| = 1\}$$

as a homogeneous space. Perhaps the most common one is to consider the orthogonal group $G = O(n+1)$ which keeps S^n invariant and its subgroup H of the maps fixing a given point $o \in S^n$, isomorphic to $O(n)$. Clearly $S^n = O(n+1)/O(n)$. The Lie algebras $\mathfrak{g} = \mathrm{Lie}\,G$ and $\mathfrak{h} = \mathrm{Lie}\,H$ enjoy the nice matrix $(1, n)$ block structure

$$\mathfrak{g} = \begin{pmatrix} 0 & -v^{\mathrm{T}} \\ v & X \end{pmatrix}, \quad \mathfrak{h} = \begin{pmatrix} 0 & 0 \\ 0 & X \end{pmatrix}.$$

The tangent spaces are $T_x S^n = \{y \in \mathbb{R}^{n+1} \mid \langle x, y \rangle = 0\}$ and the action of G on $\mathbb{R}^{n+1}$, $x \longmapsto Ax$, preserves both S^n and TS^n, i.e., $y \longmapsto Ay$ maps $T_x S^n \longmapsto T_{Ax} S^n$. Moreover, they preserve the scalar products on the tangent spaces and thus S^n enjoys $O(n+1)$ as isometries of the natural structure of a Riemannian manifold (S^n, g).

Observation 1 There are no other isometries of S^n apart from $O(n+1)$.

The standard way to see the above observation holds true is the following. Consider a unit vector $e_1 \in \mathbb{R}^{n+1}$ and an isometry $\phi \in \mathrm{Isom}(S^n, g)$. Then $\phi(e_1) \in S^n$ and $\phi(e_1) = A(e_1)$ for some $A \in O(n+1)$. Moreover, elements of the form $A^{-1} \circ \phi$ are in the isotropy group of e_1. As well known the Riemannian isometries are (on connected components) uniquely determined by their differential in one point. Thus the latter map coincides with its differential at e_1 and we are finished.

Another possibility is based on the Maurer–Cartan form. It is more complicated, but much more conceptual.

Consider the principal H-bundle $G \xrightarrow{p} G/H$ over $S^n \cong G/H$ equipped with the Maurer–Cartan form $\omega \in \Omega^1(G, \mathfrak{g})$. Since $\mathfrak{g}$ splits as $\mathfrak{g} = \mathfrak{h} \oplus \mathfrak{n}$ (as $\mathfrak{h}$-module), the Maurer–Cartan form also splits as $\omega = \omega_{\mathfrak{h}} \oplus \omega_{\mathfrak{n}}$, where the first part is the principal connection $\omega_{\mathfrak{h}}$ on G and the second part is the soldering form $\omega_{\mathfrak{n}}$. The soldering form provides for all $A \in G$ isomorphisms

$$\omega_{\mathrm{n}}\colon T_A G / V_A G \overset{\cong}{\to} T_{p(A)} S^n \cong \mathbb{R}^n ,$$

where $V_A G := \ker \omega_{\mathrm{n}}$ is the vertical subspace. This means ω_{n} makes $G \to S^n$ into a reduction of the linear frame bundle $P^1 S^n$ to H.

Now any isometry ϕ lifts to the level of frame bundles and can be restricted to G and thus we have a lift $\tilde{\phi}\colon G \to G$ such that $\tilde{\phi}^* \omega_{\mathrm{n}} = \omega_{\mathrm{n}}$. Because $\omega_{\mathfrak{h}}$ is principal connection preserving the metric we also have $\tilde{\phi}^* \omega_{\mathfrak{h}} = \omega_{\mathfrak{h}}$. We see that $\tilde{\phi}^* \omega = \omega$, i.e., $\tilde{\phi}$ preserves the Maurer–Cartan form.

Notice that in this setting, $\omega_{\mathfrak{h}}$ must be the only torsion free metric connection on S^n. Thus, we arrived at the canonical Cartan connection on S^n in the sense of the so-called *Cartan geometry* as defined below.

For any principal fiber bundle $\mathcal{G}$ with structure group P, we shall write r^g for the principal right action of elements in P and ζ_X means the fundamental vector field, $\zeta_X(u) = \frac{d}{dt}|_0 r^{\exp tX}(u)$.

Definition 2.1 For a pair $H \subset G$ of a Lie group and its Lie subgroup, a Cartan geometry is a principal H-bundle $p : \mathcal{P} \to M$ endowed with a $\mathfrak{g}$-valued 1-form $\omega \in \Omega^1(\mathcal{P}, \mathfrak{g})$ satisfying for all $h \in H$, $X \in \mathfrak{h}$, $u \in \mathcal{P}$ the following three properties

$$\mathrm{Ad}(h^{-1}) \circ \omega = (r^h)^* \omega, \tag{2.1}$$

$$\omega(\zeta_X(u)) = X, \tag{2.2}$$

$$\omega(u)\colon T_u \mathcal{P} \overset{\cong}{\to} \mathfrak{g}. \tag{2.3}$$

Now, our observation follows from a general result:

Theorem 2.1 (Fundamental Theorem of Calculus) *Let ω_G be the Maurer–Cartan form of a Lie group G with the Lie algebra $\mathfrak{g}$, M a smooth manifold endowed with a 1-form $\omega \in \Omega^1(M, \mathfrak{g})$. Then for each $x \in M$ there is a neighborhood $U \ni x$ and $f\colon U \to G$ such that $f^* \omega_G = \omega$, if and only if*

$$d\omega + \frac{1}{2}[\omega, \omega] = 0. \tag{2.4}$$

If M is connected and $f_1, f_2\colon M \to G$ with $f_1^* \omega_G = f_2^* \omega_G$ on M, then there exists a unique $c \in G$ such that $f_2 = c f_1$ on M.

Under the additional requirement that $Tf : T_x M \to T_{f(x)} G$ is a linear isomorphism for each point x, the theorem shows that the local Lie group structure is uniquely determined by the Maurer–Cartan form satisfying (2.4).

The theorem is proved by building the graph of the mapping f, see [11, section 1.2.4]. Notice, in dimension one the condition is empty and so with the additive group $G = \mathbb{R}$ we obtain just the existence of primitive functions up to a constant. If G was the multiplicative group $\mathbb{R}_+$, the theorem would show how the logarithmic derivatives prescribe the functions, up to a constant multiple.

In our case each isomorphism $\phi : S^n \to S^n$ lifts to the unique map $f : G \to G$ satisfying $f^*\omega = \omega$. The Maurer–Cartan forms on all Lie groups satisfy the condition in (2.4) and thus, even locally, f can differ from the identity map only by an element of G.

2.1.2   Conformal Riemannian Sphere

A conformal Riemannian manifold $(M, [g])$ is a manifold M with a conformal class of metrics. Two metrics g, $\tilde{g}$ are representatives of the same conformal class if they differ by some positive function, $\tilde{g} = \Omega^2 g$, $\Omega \in C^\infty(M)$. Conformal isometry is a diffeomorphism, whose differentials at all points belong to the conformal orthogonal group $CO(n)$ for the given structures on the tangent spaces.

The conformal sphere is $(S^n, [g])$ where $[g]$ includes the standard round metric. Let us discuss the following question: *What is the group of all conformal isomorphisms on S^n making it into a homogeneous space G/P?*

Option 1 We can go the "brutal force" way. Take $\mathbb{R}^n$ with the conformal class containing the Euclidean metric and write down the PDEs for an arbitrary locally defined conformal isomorphism ϕ, i.e., we request the differentials of ϕ are in the Lie algebra $co(n)$ at all points. There is the famous Liouville theorem saying that each such ϕ is generated by the Euclidean motions, the dilations, and the sphere inversions. An elementary (but tricky) proof can be found in [20, section 5.4]. In particular, if we compactify $\mathbb{R}^n$ by the one point at infinity, we can extend all such local diffeomorphisms to globally defined conformal maps on S^n.

Let us try to do it in a smart way. Consider $\mathbb{R}^{n+2}$ with the pseudo-Euclidean metric $Q(x, x) = 2x_0 x_{n+1} + x_1^2 + \cdots + x_n^2$ of signature $(n + 1, 1)$ and define C to be the null-cone of this metric. Now, we may identify the sphere S^n with the projectivization $\mathbb{P}C$ of this cone and write down the action of all the latter maps in projective coordinates on $\mathbb{P}\mathbb{R}^{n+2}$. We can represent the null-vectors of the affine $\mathbb{R}^n \subset S^n$ as $(1 : x : -\frac{1}{2}\|x\|^2)$, while the remaining infinite point in S^n as $(0 : 0 : 1)$. Now we may easily identify the above conformal maps as actions of particular matrices in $O(n + 1, 1)$ on the projectivized cone $\mathbb{P}C$.

$$\begin{pmatrix} 1 \\ x \\ -\frac{1}{2}\|x\|^2 \end{pmatrix} \longmapsto \begin{pmatrix} a^{-1} & 0 & 0 \\ 0 & A & 0 \\ 0 & 0 & a \end{pmatrix} \begin{pmatrix} 1 \\ x \\ -\frac{1}{2}\|x\|^2 \end{pmatrix} \tag{2.5}$$

$$\begin{pmatrix} 1 \\ x \\ -\frac{1}{2}\|x\|^2 \end{pmatrix} \longmapsto \begin{pmatrix} 1 & 0 & 0 \\ v & E & 0 \\ -\frac{1}{2}\|v\|^2 & -v^T & 1 \end{pmatrix} \begin{pmatrix} 1 \\ x \\ -\frac{1}{2}\|x\|^2 \end{pmatrix} \tag{2.6}$$

$$\begin{pmatrix} 1 \\ x \\ -\frac{1}{2}\|x\|^2 \end{pmatrix} \longmapsto \begin{pmatrix} 0 & 0 & -2 \\ 0 & E & 0 \\ -\frac{1}{2} & 0 & 0 \end{pmatrix} \begin{pmatrix} 1 \\ x \\ -\frac{1}{2}\|x\|^2 \end{pmatrix} \tag{2.7}$$

Notice in the last line that the sphere inversion $\sigma \in O(n+1, 1)$ is in a different component than the unit, while the non-trivial maps fixing the origin and having the identity as differential there are obtained by composing $\sigma \circ \tau_v \circ \sigma$, with the translation $\tau_v \in O(n+1, 1)$ from (2.6), cf. [20, section 5.10].

Option 2 Similarly to the Riemannian case, we first choose the right homogeneous space $S^n = G/P$ with $G = O(n+1, 1)$, P the isotropy group of one fixed point in S^n, and show that G is just the group of all conformal isomorphisms. Again, we can achieve that by building a reasonably normalized Cartan geometry for each conformal Riemannian manifold. Then the Maurer–Cartan form ω_G of G will be preserved by all conformal morphisms and thus the Theorem 2.1 applies.

We shall come back to such normalizations of Cartan geometries later in the fifth lecture.

At the level of Lie algebras, $\mathfrak{g} = \text{Lie } G$ decomposes as $\mathfrak{g} = \mathfrak{g}_{-1} \oplus \mathfrak{p}$, where $\mathfrak{g}_{-1}$ are the infinitesimal translations with matrices

$$\mathfrak{g}_{-1} = \left\{ \begin{pmatrix} 0 & 0 & 0 \\ v & 0 & 0 \\ 0 & -v^T & 0 \end{pmatrix} \right\}$$

while

$$\mathfrak{p} = \mathfrak{g}_0 \oplus \mathfrak{g}_1 = \left\{ \underbrace{\begin{pmatrix} -a & 0 & 0 \\ 0 & A & 0 \\ 0 & 0 & a \end{pmatrix}}_{\mathfrak{co}(n)} \oplus \underbrace{\begin{pmatrix} 0 & w & 0 \\ 0 & 0 & -w^T \\ 0 & 0 & 0 \end{pmatrix}}_{\mathbb{R}^n} \right\}.$$

Clearly, $\mathfrak{g}_1$ is a $\mathfrak{p}$-submodule (actually an ideal), while $\mathfrak{g}_0$ is identified with the $\mathfrak{p}$-module $\mathfrak{p}/\mathfrak{g}_1$ with the trivial action of $\mathfrak{g}_1$. This is the well known decomposition of $\mathfrak{p}$ into the reductive quotient and the nilpotent submodule.

At the level of Lie groups, this corresponds to the splitting of the isotropy group P into the semi-direct product of $CO(n)$ containing the conformal isomorphisms fixing the origin and determined by their first derivatives, and the nilpotent normal subgroup $P_+ \subset P$ of those conformal isomorphisms fixing the origin with trivial first differential and determined by the second order derivatives. We shall also write $G_0 = P/P_+$ and this reductive group decomposes further into the semisimple part $O(n)$ and the center $\mathbb{R} \setminus \{0\}$.

2.1.3   Towards Tractors

The conformal Riemannian structure on S^n can be read off the standard metric Q on $\mathbb{R}^{n+2}$ as follows. Any choice of a non-zero section of the null-cone C (e.g., we may choose one of the non-zero components C_+ of C and consider sections there), seen as line bundle over its projectivization S^n, provides the identification of the tangent bundle $T_p S^n$ with the quotient of

$$T_p C_+ = \{z \in \mathbb{R}^{n+2} \mid Q(z, p) = 0\} = \langle p \rangle^\perp$$

by the line $\langle p \rangle$ (notice p is null, so this line is in the tangent space). Clearly $\langle p \rangle^\perp / \langle p \rangle$ is linearly isomorphic to $T_p S^n$ and since p is null, Q induces a positive definite metric on $T_p S^n$. If we multiply p by a constant $a \neq 0$, then the induced metric will change by the positive multiple a^2. By the very construction, this conformal structure is invariant with respect to the natural action of $O(n + 1, 1)$ on the cone C. Thus, we may also view C_+ as the square root of the line bundle of the conformal metrics in this class.

Of course, there is no preferred affine connection on S^n in this picture. But if we consider the flat affine connection ∇ on $\mathbb{R}^{n+2}$, then we can consider the parallel (constant) vector fields in the trivial vector bundle $C \times \mathbb{R}^{n+2}$ along the null-lines in C and view them as fields in the trivial vector bundle $\mathcal{T} S^n = S^n \times \mathbb{R}^{n+2}$.

The slight problem with this point of view is that we should expect that the fibers of $\mathcal{T} S^n$ split into the "vertical part" along the null-lines in C, the "tangent part" to S^n and the complementary 1-dimensional part in $\mathbb{R}^{n+2}$. While the vertical part is well defined, such a splitting clearly depends on the choice of the identification of S^n with a section of $C \to S^n$. Moreover, we should hope to inherit an invariant connection from the flat connection ∇ on $\mathbb{R}^{n+2}$. Before answering these questions, we are going to indicate a much simpler abstract description of such objects and we come back to these functorial objects and constructions in the fourth lecture.

Let $H \subset G$ be a Lie subgroup and $G \to G/H$ the corresponding Klein geometry. Notice that $G \to G/H$ is a principal H-bundle. Consider any linear representation $\mathbb{V}$ of G and the associated bundle $\mathcal{V} = G \times_H \mathbb{V}$, i.e., the classes of the equivalence relations on $G \times \mathbb{V}$ given by $(u, v) \sim (u \cdot h, h^{-1} \cdot v)$.[1]

In particular, we may identify the class $[\![u, v]\!]$ with the couple $(u \cdot H, u \cdot v)$. Indeed, taking another representative, we arrive at

$$((u \cdot h) \cdot H, u \cdot h \cdot (h^{-1} \cdot v)) = (u \cdot H, u \cdot v)$$

and thus $\mathcal{V}$ is the trivial bundle on M

$$\mathcal{V} = G/H \times \mathbb{V}.$$

---

[1] Here $u \cdot h$ is multiplication in G and $h^{-1} \cdot v$ is the left-action of H or G on $\mathbb{V}$ given by the chosen representation.

Moreover, there is the Maurer–Cartan form ω on G. Extending $G \to G/H$ to the principle G-bundle $\tilde{G} = G \times_H G \to G/H$, the form ω uniquely extends to a principal connection form $\tilde{\omega}$ on $\tilde{G}$. Finally, we can further identify $\mathcal{V}$ as the associated space $\mathcal{V} = \tilde{G} \times_G \mathbb{V}$. Thus we see that there is the induced connection ∇ on all such bundles $\mathcal{V}$.

2.1.4 Tracy Thomas' Tractors $\mathcal{T}$

Now we come back to the conformal sphere and we apply the above abstract construction. Thus, $G = O(n + 1, 1)$, and $H = P \subset G$ is the isotropy group of the fixed origin $(1 : 0 : 0)$, i.e., the Poincare subgroup in G with the Lie algebra $\mathfrak{p}$ as discussed above. Further, we may take $\mathbb{T} = \mathbb{R}^{n+2}$ with the standard action of G.

The final ingredients we need are the weights of line bundles or more general tensor bundles on conformal manifolds. Consider $\mathbb{R}[w]$ as the representation of P such that P_+ and $O(n)$ act trivially, while the central element $\lambda = \exp(aE) \in G_0$ acts as $\lambda \cdot x = e^{-aw} x$. Here E is the so-called grading element in $\mathfrak{g}$, i.e.,

$$\lambda = \exp \begin{pmatrix} a & 0 & 0 \\ 0 & 0 & 0 \\ 0 & 0 & -a \end{pmatrix} = \begin{pmatrix} e^a & 0 & 0 \\ 0 & \mathrm{Id}_n & 0 \\ 0 & 0 & e^{-a} \end{pmatrix}.$$

Now, the line bundles of weights w are defined as

$$\mathcal{E}[w] = G \times_P \mathbb{R}[w].$$

At the level of the infinitesimal action, the central element $aE \in \mathfrak{g}_0$ will act as $a \cdot x = -wax$. Notice the minus sign convention—this is because we want the line bundle of the conformal metrics to get the weight two.

Taking tensor products with line bundles, we arrive at the general weighted bundles $\mathcal{V}[w] = \mathcal{V} \otimes \mathcal{E}[w]$ for any representation $\mathbb{V}$ of P or even G.

Now, look at the action (2.5) on the components of $\mathbb{T}$. We see immediately that the representation space $\mathbb{T} = \mathbb{R}^{n+2}$ splits as G_0-module into

$$\mathbb{T} = \mathbb{R}[1] \oplus \mathbb{R}^n[-1] \oplus \mathbb{R}[-1],$$

where the right ends are P-submodules. In particular, $\mathbb{R}[-1]$ is a P-submodule, while $\mathbb{R}[1]$ is the projecting component. Thus, the trivial bundle $\mathcal{T}$ splits (once a section of C_+ and thus one of the metrics in the class is fixed):

$$\mathcal{T} = \mathcal{E}[1] \oplus \overset{\displaystyle VC_+}{\overbrace{T S^n[-1] \oplus \mathcal{E}[-1]}}.$$
$$\underbrace{\phantom{\mathcal{E}[1] \oplus T S^n[-1] \oplus \mathcal{E}[-1]}}_{TC_+}$$

As mentioned above, the bundle $\mathcal{T}$ comes equipped with the canonical metric induced by Q and $TC_+ = (VC_+)^\perp$. Thus, there is the positive definite metric $\mathbf{g} : TS^n[-1] \times TS^n[-1] \rightarrow \mathcal{E}$, i.e., $\mathbf{g}$ is a section of $S^2(T^*S^n)[2]$. This is the conformal class of metrics on S^n viewed as the section of a weighted metric bundle and it allows us to raise and lower tensor indices of arbitrary tensors exactly as in the Riemannian case, but at the expense of adding or subtracting the weight 2. For example, we may write $\mathcal{T} = \mathcal{E}[1] \oplus T^*S^n[1] \oplus \mathcal{E}[-1]$.

Finally, any non-zero section σ of the projecting part $\mathcal{E}[1]$ provides the Riemannian metric $g = \sigma^{-2}\mathbf{g}$.

2.2 Conformal to Einstein and the Tractor Connection

Tracy Thomas came across his conformal tractors in [22], when constructing basic invariants of conformal geometry via a linear connection on a suitable vector bundle (instead of building an absolute parallelism in the Cartan's approach). He succeeded in finding the simplest of such vector bundles, together with an invariant linear connection. He also worked out the necessary transformation properties based on the so-called Schouten tensor.

All these objects were reinvented in [1] and here the authors also discussed the following question: *Given a conformal class* [g] *on a manifold, is there a representative of the class which is an Einstein metric?* We shall follow this development and thus we shall find the Thomas' tractors when prolonging a conformally invariant geometric PDE (also following [12]).

In the sequel, we shall use the abstract index formalism. Moreover we shall mostly not distinguish between the bundles VM and the spaces of their sections $\Gamma(VM)$. Thus, we shall talk about vector fields in $\mathcal{E}^a$ or one-forms in $\mathcal{E}_a$. Similarly, η_{ab} is either a two-form in $\mathcal{E}_{ab}$ or a $\mathcal{E}_a$-valued one-form. As usual, repeated indices at different positions (lower versus upper) mean the relevant trace.

2.2.1 The Einstein Scales

Recall that the curvature $R_{ab}{}^c{}_d$ of the unique torsion free metric connection ∇ decomposes into the trace-free Weyl tensor $W_{ab}{}^c{}_d$ and the Ricci tensor R_{ab}. We shall see later why a trace-adjusted version of R_{ab}, the so-called Schouten tensor,

$$P_{ab} = \frac{1}{n-2}\left(R_{ab} - \frac{1}{2(n-1)}Rg_{ab}\right),$$

where $R = g^{ab}R_{ab}$ is the scalar curvature, is very useful. (Notice, here we use the opposite sign convention for the Schouten tensor P than in [11], i.e. it is the same as in [12].)

Of course, we should believe there is an overdetermined distinguished PDE system on the scales, i.e., the choices of the metrics in the class, whose solutions correspond to the Einstein scales. We can write down all such PDEs with the help of any of the metrics in the class and, as a matter of fact, the equation must be independent of our choice, i.e., *conformally invariant*. A straightforward check reveals (we shall come to such techniques later) that the following equation on (the square roots of) the scales σ in $\mathcal{E}[1]$ is invariant

$$\boxed{\nabla_{(a}\nabla_{b)_0}\sigma + P_{(ab)_0}\sigma = 0} \tag{2.8}$$

Now, if σ is a nowhere zero solution, we may write the equation using the metric connection corresponding to σ and thus, σ is parallel and we arrive at $P_{(ab)_0} = 0$. This is exactly the condition to be Einstein, i.e., the trace-free part of Ricci vanishes and $\sigma^{-2}\mathbf{g}_{ab}$ is Einstein.

We are going to apply the classical method of prolongation of overdetermined systems of PDEs to show that solutions of (2.8) are equivalent to *parallel tractors* in $\mathcal{T}$.

First, we add trace part $\rho\mathbf{g}$ to the Eq. (2.8), i.e., ρ is a new (-1)-weighted quantity $\rho \in \mathcal{E}[-1]$ and the new equation becomes

$$\nabla_{(a}\nabla_{b)}\sigma + P_{(ab)}\sigma + \mathbf{g}_{ab}\rho = 0. \tag{2.9}$$

Moreover, we know that the Ricci curvature is symmetric for all scales, i.e., the Levi-Civita connections of the metrics in the class, and thus the Schouten tensor is symmetric too. Finally, the antisymmetric part of the second order derivative is given by the action of the curvature $R_{ab}{}^c{}_d$ as a 2-form valued in the Lie algebra $\mathfrak{so}(n, \mathbb{R})$ and these values have no central component to act on the densities $\mathcal{E}[w]$. Thus our equation becomes

$$\nabla_a\nabla_b\sigma + P_{ab}\sigma + \mathbf{g}_{ab}\rho = 0. \tag{2.10}$$

In the next step, we give the derivative $\nabla_a\sigma$ the new name $\mu_a = \nabla_a\sigma \in \mathcal{E}_a[1]$.[2] Thus the latter Eq. (2.10) can be rewritten as the system of two first order equations

$$\boxed{\begin{aligned} \nabla_a\sigma - \mu_a &= 0 \\ \nabla_a\mu_b + P_{ab}\sigma + \mathbf{g}_{ab}\rho &= 0 \end{aligned}} \tag{2.11}$$

This system is not yet closed since there is still the uncoupled variable ρ. Thus we have to prolong the system and we need some computational preparation first.

---

[2] We know that μ_a must be of weight 1 because covariant differentiation does not alter weights and σ is already of weight 1.

Recall the invariant conformal metric $\mathbf{g}$ is covariantly constant in all scales and differentiate (2.10):

$$\nabla_a \nabla_b \nabla_c \sigma + \mathbf{g}_{bc} \nabla_a \rho + (\nabla_a P_{bc})\sigma + P_{bc} \nabla_a \sigma = 0. \tag{2.12}$$

Contract (2.12) by hitting it with $\mathbf{g}^{ab}$ and $\mathbf{g}^{bc}$, respectively:

$$\Delta(\nabla_c \sigma) + \nabla_c \rho + \nabla^a P_{ac}\sigma + P^a{}_c \nabla_a \sigma = 0, \tag{2.13}$$

$$\nabla_a(\Delta \sigma) + n\nabla_a \rho + \nabla_a P \sigma + P \nabla_a \sigma = 0, \tag{2.14}$$

where P is the trace of P_{ab}. Next, contracting the Bianchi identity and some straightforward computations lead to

$$\nabla^a P_{ac} = \nabla_c P \tag{2.15}$$

$$[\nabla_c, \Delta] = R_{cb}{}^b{}_d \nabla^d. \tag{2.16}$$

Subtracting (2.13) from (2.14) and using (2.15) and (2.16) we arrive at

$$(n-1)\nabla_c \rho + P\nabla_c \sigma - P^a{}_c \nabla_a \sigma + R_{cb}{}^b{}_d \nabla^d \sigma = 0. \tag{2.17}$$

Further notice

$$R_{cb}{}^b{}_d \nabla^d \sigma = -R_{ca} \nabla^a \sigma = (2-n)P_c{}^a \nabla_a \sigma - \nabla_c P \sigma$$

which together with (2.17) yields, up to the constant factor $n-1$

$$\boxed{\nabla_c \rho - P_c{}^a \mu_a = 0} \tag{2.18}$$

and our system of equations closes up. Summarizing, the Einstein scales correspond to nowhere zero solutions of our system of three first order equations coupling σ, μ_a, and ρ and all this should be understood in terms of conformally invariant objects and operations.

Indeed, this is the content of the following theorem. For now, we formulate it only for the solutions to our equations on the sphere S^n, although our discussion on the equations has concerned general conformal Riemannian manifolds.

Theorem 2.2 *Let $\mathcal{T} = \mathcal{E}[1] \oplus T^*S^n[1] \oplus \mathcal{E}[-1]$ be the bundle of the Thomas' tractors on the conformal sphere. Define the following operator on $\mathcal{T}$*

$$\nabla_a^{\mathcal{T}} \begin{pmatrix} \sigma \\ \mu_a \\ \rho \end{pmatrix} = \begin{pmatrix} \nabla_a \sigma - \mu_a \\ \nabla_a \mu_b + \mathbf{g}_{ab}\rho + P_{ab}\sigma \\ \nabla_a \rho - P_{ab}\mu^b \end{pmatrix}, \tag{2.19}$$

where ∇_a on the right-hand side refers to the Levi-Civita connection of the metric $\sigma^{-2}\mathbf{g}$.

The operator $\nabla^{\mathcal{T}}$ is a linear connection on $\mathcal{T}$ which is conformally invariant. Moreover, solutions to (2.8) are in bijective correspondence with parallel tractors, i.e., with sections $t \in \Gamma(\mathcal{T})$ such that $\nabla^{\mathcal{T}} t = 0$.

Notice that on the sphere, $\mathcal{T}$ is the trivial bundle $S^n \times \mathbb{T}$ and we shall see soon that $\nabla^{\mathcal{T}}$ is the flat connection there which we mentioned earlier. So we also postpone the proof of this theorem. (Actually, we see immediately that this is a connection, but we should check its curvature and, in particular, how it depends on the choice of the fixed metric.)

Obviously, all the parallel tractors are determined uniquely by their values in $\mathbb{T}$ in the origin. In particular, we managed to compute all Einstein metrics on the conformal sphere.

2.2.2 Conformal Invariance

What do we really mean when saying that objects or operations are *conformally invariant*?

The intuitively obvious answer should be that they are independent of our choice of the metric in the conformal class. So we should start to look how the covariant derivative changes if we change the scale. Consider the change of our metric by taking $\hat{g} = \Omega^2 g$ with a positive smooth function Ω, and write $\Upsilon_a = \Omega^{-1}\nabla_a\Omega$.

Lemma 2.1 *Let $\hat{\nabla}$ be the Levi-Civita connection for the rescaled metric $\hat{g}$. Then for all $v \in \mathcal{E}^a$, $\alpha \in \mathcal{E}_a$, $\rho \in \mathcal{E}[w]$*

$$\hat{\nabla}_a v^b = \nabla_a v^b + \Upsilon_a v^b - \Upsilon^b v_a + \Upsilon^c v_c \delta_a^b \tag{2.20}$$

$$\hat{\nabla}_a \alpha_b = \nabla_a \alpha_b - \Upsilon_a \alpha_b - \Upsilon_b \alpha_a + \Upsilon^c \alpha_c g_{ab} \tag{2.21}$$

$$\hat{\nabla}_a \rho = \nabla_a \rho + w\Upsilon_a. \tag{2.22}$$

Proof Recall the Christoffel symbols of Levi-Civita connection are expressed in any coordinates via the derivative of the metric coefficients (we write ∇_i for the partial derivatives here)

$$\Gamma^i{}_{jk} - \frac{1}{2}g^{i\ell}(\nabla_k g_{\ell j} + \nabla_j g_{\ell k} - \nabla_\ell g_{jk}). \tag{2.23}$$

Conformal rescaling of the metric $g \longmapsto \hat{g} = \Omega^2 g$ affects all other objects derived from metric, e.g., the new inverse metric is $\hat{g}^{-1} = \Omega^{-2}g^{-1}$. Thus, the Christoffels (2.23) change

$$\hat{\Gamma}^i{}_{jk} = \Gamma^i{}_{jk} + \frac{1}{\Omega}\left(\delta^i_j \nabla_k \Omega + \delta^i_k \nabla_j \Omega - g_{jk}\nabla^i \Omega\right)$$
$$= \Gamma^i{}_{jk} + \delta^i_j \Upsilon_k + \delta^i_k \Upsilon_j - g_{jk}\Upsilon^i. \tag{2.24}$$

Now recall, the covariant derivative is in coordinates given as the directional derivative modified by the action of the Christoffels viewed as $\mathfrak{o}(n)$-valued one-form. Thus the latter formula provides exactly the three formulae in the statement.

□

The formulae from the lemma allow to compute easily the changes of conformal derivatives on all weighted tensor bundles.

For example, considering possible first order operators on weighted forms $\mathcal{E}_a[w]$, we get

$$\hat{\nabla}_a \alpha_b = \nabla_a \alpha_b + (w-1)\Upsilon_a \alpha_b - \Upsilon_b \alpha_a + \Upsilon^c \alpha_c g_{ab},$$

and we immediately see that the antisymmetric part $\nabla_{[a}\alpha_{b]}$ is invariant for the weight $w = 0$ (this is the exterior differential on one-forms), the trace-free part of the symmetrization $\nabla_{(a}\alpha_{b)_0}$ is invariant for $w = 2$ (we may view this as the operator on the vector fields in $\mathcal{E}^a = \mathcal{E}_a[2]$ and the kernel describes the conformal Killing vector fields), and finally the trace $\nabla^a \alpha_a$ is invariant for $w = 2 - n$ (this is the divergence of vector fields with weight $-n$).

Let us look at the geometric objects next. On the conformal sphere S^n, the category of natural objects was defined in 2.1.3—those are the homogeneous bundles $G \times_P \mathbb{V}$ corresponding to any representation of P. If the representation comes from a representation of $G_0 = \mathrm{CO}(n)$, extended by the trivial representation of $P_+ = \exp \mathfrak{g}_1$, the corresponding bundles extend to all conformal Riemannian manifolds. Indeed, since general conformal Riemannian manifolds are given as reduction of the linear frame bundles to the structure group G_0, such bundles are well defined on all of them.

If we deal with more general P-representations, then we arrive at sums of the latter bundles as soon as we fix a metric g in the conformal class, but the components are not given invariantly. We shall explain the general procedure in the next lecture and, in particular, we shall see how this behavior extends and defines such bundles on all conformal Riemannian manifolds. For now, just believe that in the case of the Thomas' tractor bundles we face the following transformation rule

$$\begin{pmatrix} \hat{\sigma} \\ \hat{\mu}_a \\ \hat{\rho} \end{pmatrix} = \begin{pmatrix} \sigma \\ \mu_a + \sigma\Upsilon_a \\ \rho - \Upsilon^c \mu_c - \frac{1}{2}\Upsilon^c \Upsilon_c \sigma \end{pmatrix} \tag{2.25}$$

Of course, a straightforward (and really tedious) computation can reveal that considering the formulae for the transformations of covariant derivatives from the above Lemma and (2.25), the linear tractor connection ∇^T_a is a well defined conformally invariant operator on the tractors. Fortunately, we do not have to check this the pedestrian way and can wait for general reasons.

2.3   Parabolic Geometries

We met the general Cartan geometries with the model G/H in the Definition 2.1. If the Lie group G is semisimple and the subgroup H is a parabolic subgroup in G, we talk about the *parabolic geometries*. This class of Cartan geometries includes many very important examples and provides a unified theory for all of them. In this lecture we shall introduce some basic features and clarify many phenomena in the conformal case on the way. Detailed exposition of the background, including the necessary representation theory, is available in [11].

In general, the definition of the parabolic subgroups is a little subtle. For us, the simplest approach is via their Lie algebras. The parabolic ones are those which contain a Borel subalgebra and the choices of parabolic subalgebras $\mathfrak{p} \subset \mathfrak{g}$ correspond to graded decompositions of the semisimple Lie algebras

$$\mathfrak{g} = \mathfrak{g}_{-k} \oplus \cdots \oplus \mathfrak{g}_k.$$

This means that Lie brackets respect the grading, $[\mathfrak{g}_i, \mathfrak{g}_j] \subset \mathfrak{g}_{i+j}$, and $\mathfrak{p} = \mathfrak{g}_0 \oplus \mathfrak{p}_+ = \mathfrak{g}_0 \oplus \mathfrak{g}_1 \oplus \cdots \oplus \mathfrak{g}_k$ is the decomposition into the reductive quotient $\mathfrak{g}_0$ and nilpotent subalgebra $\mathfrak{p}_+$. Moreover, there always is the unique *grading element E* in the center of $\mathfrak{g}_0$ with the property $[E, X] = jX$ for all $X \in \mathfrak{g}_j$.

The closed Lie subgroups $P \subset G$ are called parabolic if their algebras $\mathfrak{p} = \operatorname{Lie} P$ are parabolic.

If G is a complex semisimple Lie subgroup, then there is a nice geometric description: $P \subset G$ is parabolic if and only if G/P is a compact manifold (and then it is a compact Kähler projective variety), see, e.g., [24, Section 1.2]. In the real setting, the so-called generalized flag varieties G/P with parabolic P are always compact.

2.3.1 |1|-graded Parabolic Geometries

For the sake of simplicity, we shall restrict ourselves to the so-called |1|-graded cases here, i e , $k = 1$. Thus we shall deal with Lie groups with the algebras

$$\mathfrak{g} = \mathfrak{g}_{-1} \oplus \underbrace{\mathfrak{g}_0 \oplus \mathfrak{g}_1}_{\mathfrak{p}}, \tag{2.26}$$

where $\mathfrak{p}$ refers to the parabolic subalgebra.

Now, we consider Cartan geometries modelled on $G \to G/P$, i.e., principal P-bundles $\mathcal{G} \to M$ with Cartan connections $\omega \in \Omega^1(\mathcal{G}, \mathfrak{p})$. Such a connection ω splits due to (2.26) as

$$\omega = \omega_{-1} \oplus \omega_0 \oplus \omega_1.$$

We shall further consider all reductions of the principal bundles $\mathcal{G}$ to the structure group $G_0 = P/\exp \mathfrak{g}_1$, i.e., we are interested in all equivariant mappings

$$\sigma : \mathcal{G}_0 = \mathcal{G}/\exp \mathfrak{g}_1 \to \mathcal{G}$$

with respect to the right principal actions. The diagram below summarizes our situation (notice we are also fixing the subgroup G_0 in the semi-direct product $P = G_0 \ltimes \exp \mathfrak{g}_1$, following the splitting of the Lie algebra)

$$G_0 \subset P \curvearrowright \mathcal{G} \underset{\sigma}{\overset{}{\rightleftarrows}} \mathcal{G}_0 \longrightarrow M.$$

The Cartan connection ω allows us to identify the cotangent bundle T^*M with $\mathcal{G} \times_P \mathfrak{g}_1$, where the action of P_+ is trivial. Similarly $TM \simeq \mathcal{G} \times_P \mathfrak{g}/\mathfrak{p}$, where $\mathfrak{g}/\mathfrak{p} \simeq \mathfrak{g}_{-1}$, again with trivial action of P_+. The duality is provided by the Killing form on $\mathfrak{g}$.

Recall that all sections ϕ of associated bundles $\mathcal{G} \times_P \mathbb{V}$ are identified with equivariant functions $f : \mathcal{G} \to \mathbb{V}$, i.e., $\phi(x) = [\![u(x), f(u(x))]\!]$ and so

$$f(u \cdot p) = p^{-1} \cdot f(u).$$

Once we restrict the structure group to the reductive part of P, the pullback of the Cartan connection along σ splits into the so-called soldering form valued in $\mathfrak{g}_{-1}$, principal connection form valued in $\mathfrak{g}_0$, and the one-form valued one-form P (which we shall see as the general analog of the Schouten tensor P_{ab} from conformal geometry)

$$\sigma^*\omega = \theta \oplus \sigma^*\omega_0 \oplus \mathsf{P}.$$

We can also take the other way round—since P_+ is contractible (as the exponential image of a nilpotent algebra), we may start with $\mathcal{G} = \mathcal{G}_0 \times P_+$, fix one of the (reasonably normalized) pullbacks $\sigma^*\omega_0$ and use some suitable P to define the Cartan connection on the entire $\mathcal{G}$. We shall see later, the Schouten tensor (with the opposite sign, see the comment in the beginning of the second lecture) is the right choice to get the normalized Cartan connection in the case of conformal Riemannian geometries, taking one of the Levi-Civita connections for $\sigma^*\omega_0$. But we shall stay at the level of general Cartan connections now, so P is just the relevant pullback.

Two such reductions differ by a one-form Υ, viewed as equivariant function $\Upsilon : \mathcal{G}_0 \to \mathfrak{g}_1$:

$$\hat{\sigma} = \sigma \cdot \exp \Upsilon.$$

2.3.2 Natural Bundles and Weyl Connections

For each representation $\mathbb{V}$ of P there is the functorial construction of the bundles $\mathcal{V} = \mathcal{G} \times_P \mathbb{V}$ and the morphisms of the Cartan geometries act on them in the obvious way.

Actually we do not need the Cartan connection for this definition, but notice that the morphisms of the principal bundles respecting the Cartan connections are rather rigid in the following sense. If we fix their projection to the base manifolds, the freedom in covering them is described by the kernel K of the homogeneous models, i.e., the subgroup of P acting trivially on G/P, see [11, section 1.5.3]. This means two such morphisms may differ only by right principal action of elements from K. Usually K is trivial or discrete.

Consider now a representation $\mathbb{V}$ of P and its decomposition as a G_0-module. The action of the grading element $E \in \mathfrak{g}_0$ provides the splitting

$$\mathbb{V} = \mathbb{V}_0 \oplus \cdots \oplus \mathbb{V}_k,$$

where the action of $\mathfrak{g}_1$ moves elements from $\mathbb{V}_i$ to $\mathbb{V}_{i+1}$. Clearly, any section v of $\mathcal{V}$ decomposes into the components $v_i : \mathcal{G}_0 \to \mathbb{V}_i$ as soon as we fix our reduction.

Further, fixing our reduction σ we have the affine connection $\omega_\sigma = \sigma^*(\omega_{\leqslant 0})$ (which is a Cartan connection, i.e., an absolute parallelism, on the linear frame bundle obtained as the sum of the soldering form and connection form). As well known, the corresponding covariant derivative is obtained via the constant vector fields:

$$\nabla^\sigma_\xi v(u) = \omega_\sigma^{-1}(X) \cdot v(u),$$

where $X \in \mathfrak{g}_{-1}$ corresponds to the vector ξ in a frame $u \in \mathcal{G}_0$, i.e., we simply differentiate a function in the direction of a vector (the horizontal lift of ξ to $\mathcal{G}_0$). Notice that this connection ∇^σ always respects the decomposition of $\mathcal{V}$ given by the same reduction. We call all these connections the *Weyl connections* (and we obtain the genuine Weyl connections in the conformal case with the Schouten tensor $-\mathsf{P}$, i.e., all the torsion free connections preserving the conformal Riemannian structure).

Our next theorem says, how the splitting of $\mathbb{V}$, the covariant derivative, and also the one-form P change if we change the reduction σ.

In order to formulate the results, let us introduce some further conventions. Recall, tangent vectors $\xi \in T_x M$ can be identified with right-equivariant functions X on the frames over x valued in $\mathfrak{g}_{-1} = \mathfrak{g}/\mathfrak{p}$. This identification can be written down with the help of the Cartan connection, $X = \omega_{-1}(u)(\tilde{\xi})$ for any lift $\tilde{\xi}$ of ξ to $T_u \mathcal{G}$. By abuse of notation we shall write the same symbol ξ for the vector in TM and the corresponding element X in $\mathfrak{g}_{-1}$. Similarly we shall deal with the one-forms

Υ represented by elements in $\mathfrak{g}_1$, and also the endomorphisms of TM represented by elements in $\mathfrak{g}_0$.

For instance, $\mathrm{ad}\,\Upsilon(\xi) \cdot v$ means we take the Lie algebra valued functions Υ and ξ, take the Lie bracket of their values and act by the result on the value of the function v via the representation of $\mathfrak{g}_0$ in question. Of course, we may use only such operations which ensure the necessary equivariance (which is guaranteed when taking the adjoint action within the Lie algebra).

Theorem 2.3 *Consider* $\hat{\sigma} = \sigma \cdot \exp\Upsilon$ *and use the hat to indicate all the transformed quantities. For every section* $v = v_0 \oplus \cdots \oplus v_k$ *in the representation space of the representation* $\lambda : \mathfrak{p} \to \mathfrak{gl}(\mathbb{V})$, *and vector* ξ *in the tangent bundle,*

$$\hat{v}_\ell = (\lambda(\exp(-\Upsilon))(v))_\ell = \sum_{i+j=\ell} \frac{(-1)^i}{i!} \lambda(\Upsilon)^i(v_j). \tag{2.27}$$

If λ *is a completely reducible P-representation, then*

$$\hat{\nabla}_\xi v = \nabla_\xi v - \mathrm{ad}\,\Upsilon(\xi) \cdot v. \tag{2.28}$$

Finally, the one-form ρ *transforms*

$$\hat{\rho} = \rho(\xi) + \nabla_\xi \Upsilon + \frac{1}{2}(\mathrm{ad}\,\Upsilon)^2(\xi). \tag{2.29}$$

Proof The formula (2.27) is just a direct consequence of our definitions and reflects the fact that by changing the reduction σ, the equivariant function $v : \mathcal{G} \to \mathbb{V}$ is restricted to another subset, shifted by the right action of $\exp(\Upsilon)$. Thus, the values have to get corrected by the action of $(\exp\Upsilon)^{-1} = \exp(-\Upsilon)$. The formula then follows by collecting the terms with the right homogeneities.

The transformation of the derivative is also not too difficult. Consider a section $v : \mathcal{G}_0 \to \mathbb{V}$ and recall $\nabla_\xi v$ is given with the help of any lift $\tilde{\xi}$ to $\mathcal{G}_0$:

$$\nabla_\xi v = \tilde{\xi} \cdot v(u) - \omega_0(T_u\sigma \cdot \tilde{\xi}) \cdot v(u). \tag{2.30}$$

Writing $r^p = r(\ ,p)$ and $r_u = r(u,\ )$ for the right action,

$$T_u\hat{\sigma} \cdot \tilde{\xi} = T_{\sigma(u)}r^{\exp\Upsilon(u)} \cdot T_u\sigma \cdot \tilde{\xi} + T_{\exp\Upsilon(u)}r_{\sigma(u)} \cdot T_u\exp\Upsilon \cdot \tilde{\xi}. \tag{2.31}$$

The second term in (2.31) is vertical in $\mathcal{G} \to \mathcal{G}_0$ and thus

$$\omega_0(T_u\hat{\sigma} \cdot \tilde{\xi}) = \omega_0(T_{\sigma(u)}r^{\exp\Upsilon(u)} \cdot T_u\sigma \cdot \tilde{\xi}).$$

By equivariancy of the Cartan connection ω, this equals to the $\mathfrak{g}_0$ component of $\mathrm{Ad}((\exp\Upsilon(u))^{-1})(\omega(T_u\sigma \cdot \tilde{\xi}))$. Now, notice $\omega_{-1}(T_u\sigma \cdot \tilde{\xi})$ is exactly the coordinate function representing the vector ξ. Thus, the only $\mathfrak{g}_0$ component of the latter expression is $\mathrm{ad}(-\Upsilon(u))(\xi)$ and this has to act on v in our transformation formula.

The transformation of the P tensor is also deduced from (2.31), but it is more technical and we refer to the detailed proof in [11, section 5.1.8]. $\qquad\qquad\square$

Similar formulae are available for general parabolic geometries and their Weyl connections. Just the non-trivial gradings of TM and T^*M make them much more complicated. The complete exposition can be read from [11, sections 5.1.5 through 5.1.9].

Notice also that we are allowing all reductions σ. But some of them are nicer than others—we may reduce the structure group to the semisimple part G_0^{ss} of G_0. These further reductions correspond to sections of the line bundle $\mathcal{L} = \mathcal{G}_0/G_0^{ss}$, which can be viewed as the associated bundle $\mathcal{G}_0 \times_{G_0} \exp\{wE\}$ carrying the natural structure of a principal bundle with structure group $\mathbb{R}_+$. This is the line bundle of scales and its sections correspond to Weyl connections inducing flat connections on $\mathcal{L}$. In the conformal case, these are just the choices of metrics in the conformal class. The induced connection on $\mathcal{L}$ has got the antisymmetric part of P as its curvature and thus, we can recognize such more special reductions by the fact that for these the Rho-tensor is symmetric.

2.3.3 Higher Order Derivatives

Notice, in Theorem 2.3 we provided the formula for the change of the Weyl connections for completely reducible P-modules only. This is because the formulae get very nasty for modules with non-trivial $\mathfrak{g}_1$ actions. But even dealing with tensorial bundles, iterating the derivatives always leads to such modules.

In order to avoid at least part of these hassles, we should seek for better linear connections related to our reductions σ and the fixed Cartan connection ω. An obvious choice seems to be the following one. Fixing a reduction σ consider the principal connections $\mathcal{G}$ with the connection form $\gamma^\sigma \in \Omega^1(\mathcal{G}, \mathfrak{p})$,

$$\gamma^\sigma(\sigma(u) \cdot g)(\xi) = \omega_\mathfrak{p}(\sigma(u))(Tr^{g^{-1}} \cdot \xi)$$

for all $u \in \mathcal{G}_0, \xi \in T_{\sigma(u)\cdot g}, g \in P_+$. In other words, we restrict the $\mathfrak{p}$-component of ω to the image of σ and extend it the unique way to a principal connection form.

Clearly, this connection form defines the associated linear connections on all natural bundles, we call them the *Rho-corrected Weyl connections* ∇^P. They were perhaps first introduced in [21] and exploited properly in [6]. In the case of conformal Riemannian structures, these concepts are closely related to the so-called Wünsch's conformal calculus, cf. [23].

Theorem 2.4 *Consider natural bundle $\mathcal{V} = \mathcal{G} \times_P \mathbb{V}$ and a reduction with the Weyl connection ∇ and its Rho-corrected derivative ∇^P. Then*

$$\nabla_\xi^P v = \nabla_\xi v + P(\xi) \cdot v \tag{2.32}$$

$$\hat{\nabla}_\xi^P v = \nabla_\xi^P v + \sum_{i \geq 1} \frac{(-1)^i}{i!} (\text{ad } \Upsilon)^i (\xi) \cdot v. \tag{2.33}$$

Proof In order to see the difference between ∇ and ∇^P, we can inspect the expression (2.30) with the choice of the horizontal vector field $\tilde{\xi}$ lifting ξ. Thus $\nabla_\xi v = \tilde{\xi} \cdot v$. On the other hand, choosing the lift $T\sigma \cdot \tilde{\xi}$ on $\mathcal{G}$, we obtain $\omega_0(T\sigma \cdot \tilde{\xi}) = 0$ and $\omega_1(T\sigma \cdot \tilde{\xi})$ represents $P(\xi)$. Equivariancy of v then implies our formula (2.32) along the entire image of σ.

Let us now consider the horizontal lift $\tilde{\xi}$ of ξ on $\mathcal{G}$ with respect to γ^σ. Then $\nabla_\xi^P v$ is represented by $\tilde{\xi} \cdot v$, while $\tilde{\xi} \cdot v + \gamma^{\hat{\sigma}}(\tilde{\xi}) \cdot v$ represents $\hat{\nabla}_\xi v$. By the very definition, $\omega(\sigma(u))(\tilde{\xi}) \in \mathfrak{g}_{-1}$. Thus,

$$\gamma^{\hat{\sigma}}(\sigma(u))(\tilde{\xi}) = \omega_\mathfrak{p}(Tr^{\exp \Upsilon(u)} \cdot \tilde{\xi}(\sigma(u))),$$

which is just the $\mathfrak{p}$-component of $\text{Ad}((\exp \Upsilon(u))^{-1})(\omega(\sigma(u))(\tilde{\xi}))$. Now, notice that $\omega(\sigma(u))(\tilde{\xi})$ represents ξ by values in $\mathfrak{g}_{-1}$ and the requested formula follows.   □

We should notice that the Weyl connections and the Rho corrected ones coincide on bundles coming from representations with trivial action of P_+. Of course, the transformation formulae coincide in this case, too.

2.3.4 A Few Examples

We shall go through a few homogeneous models and comment on the general "curved" situations. In all cases the actual geometric structures are given by the reductions of the linear frame bundles and the construction of the right Cartan geometry is a separate issue. We shall come back to this in the fifth lecture and work with the general choices of the Cartan connections ω here.

Conformal Riemannian Geometry The relevant Cartan geometry can be modelled by the choice $G = O(n + 1, 1)$ (there is some freedom in the choice of the group with the given graded Lie algebra $\mathfrak{g}$) and the parabolic subgroup P as we saw in detail in the first lecture.

It is a simple exercise now to recover the formulae from Lemma 2.1 by computing the brackets in the Lie algebra. In our conventions using the coordinate functions instead of fields, we can rewrite them as (notice α is valued in $\mathfrak{g}_1$, while η, ξ have got values in $\mathfrak{g}_{-1}$, and s sits in $\mathbb{R}[w]$)

$$\hat{\nabla}_\xi \eta = \nabla_\xi \eta - [\Upsilon, \xi] \cdot \eta \tag{2.34}$$

$$\hat{\nabla}_\xi \alpha = \nabla_\xi \alpha - [\Upsilon, \xi] \cdot \alpha \tag{2.35}$$

$$\hat{\nabla}_\xi s = \nabla_\xi s - (-\Upsilon(\xi))ws, \tag{2.36}$$

where we have picked up just the central component of the bracket in the last line, viewed as the multiple of the grading element.

Another, but still much more tedious exercise would be to check the conformal invariance of the tractor connection on $\mathcal{T}$. We shall develop much better tools for that in the next lecture.

We shall also enjoy much better tools to discuss second or higher order operators. For example, considering second order operators on densities $s \in \mathcal{E}[w]$, we may iterate the Rho-corrected derivative to obtain

$$\mathbf{g}^{ab}\nabla^{\mathsf{P}}_a\nabla^{\mathsf{P}}_b s = \nabla^a\nabla_a s - w\mathbf{g}^{ab}\mathsf{P}_{ab}s$$

and check that this gets an invariant operator for $w = 1 - \frac{n}{2}$, which is the famous conformally invariant Laplacian, the so-called Yamabe operator

$$Y : \mathcal{E}[1 - \frac{n}{2}] \to \mathcal{E}[-1 - \frac{n}{2}].$$

Projective Geometry   The choice of the homogeneous model is obtained from the algebra of trace-free real matrices $\mathfrak{g} = \mathfrak{sl}(n + 1, \mathbb{R})$ with the grading

$$\left(\begin{array}{c|c} \mathfrak{z} & \mathbb{R}^{n*} \\ \hline \mathbb{R}^n & \mathfrak{gl}(n, \mathbb{R}) \end{array} \right) \begin{array}{c} 1 \\ n \end{array}$$

Here $\mathfrak{z} = \mathbb{R}$ is the center, the grading element E corresponds to $\frac{n}{n+1}$ and $-\frac{1}{n+1}\,\mathrm{id}_{\mathbb{R}^n}$ on the diagonal. We may take $G = \mathrm{SL}(n + 1, \mathbb{R})$ and P the subgroup of block upper triangular matrices. The homogeneous model is then the real projective space $G/P = \mathbb{RP}^n$. On the homogeneous model, the Weyl connections transform as

$$\hat{\nabla}_\xi \eta = \nabla_\xi \eta + \Upsilon(\eta)\xi + \Upsilon(\xi)\eta$$

and so they clearly share the geodesics.

For general projective structures on manifolds M, the space of Weyl connections has to be chosen as a class of all affine connections sharing geodesics with a given one and they transform then the same way. We shall see that projective geometries are rare exceptions of parabolic geometries not given by a first order structure on the manifold.

The analog of the Thomas' tractors is the natural bundle corresponding to the standard representation of $\mathrm{SL}(n+1, \mathbb{R})$ on $\mathbb{T} = \mathbb{R}^{n+1}$. The injecting part of $\mathcal{T}$ is the line bundle $\mathcal{T}^1$ with the action of the grading element by $\frac{n}{n+1}$. The usual convention says this is the line bundle $\mathcal{E}[-1]$. Then the projecting component is the weighted tangent bundle $TM[-1]$.

Almost Grassmannian Geometry This is essentially a continuation of the previous example. We take $G = \mathrm{SL}(p, q)$ and the splitting of the matrices into blocks of sizes p and q, say $2 \leq p \leq q$. Unlike the projective case, here the geometry is determined by reducing the structure group of the tangent bundle to $\mathbb{R} \times \mathrm{SL}(p, \mathbb{R}) \times \mathrm{SL}(q, \mathbb{R})$. This corresponds to identifying the tangent bundle with the tensor product of the auxiliary bundles $\mathcal{V}^*$ and $\mathcal{W}$ of dimensions p and q, together with the identification of their top degree forms $\Lambda^p \mathcal{V} \simeq \Lambda^q \mathcal{W}^*$.

Thus, we may use the abstract indices and write $\mathcal{V} = \mathcal{E}^A$, $\mathcal{W} = \mathcal{E}^{A'}$. Then the tangent bundle is $\mathcal{E}^{A'}_A$ and the formula for the brackets in the Lie algebra says $[[\Upsilon, \xi], \eta]^{A'}_A = -\xi^{A'}_B \Upsilon^B_{B'} \eta^{B'}_A - \xi^{B'}_A \Upsilon^B_{B'} \eta^{A'}_B$. The Weyl connections are tensor products of connections on $\mathcal{V}^*$ and $\mathcal{W}$ (but not all of them). The right formula for the change of the Weyl connections is

$$\hat{\nabla}^A_{A'} \eta^{B'}_B = \nabla^A_{A'} \eta^{B'}_B + \delta^{B'}_{A'} \Upsilon^A_{C'} \eta^{C'}_B + \delta^A_B \Upsilon^C_{A'} \eta^{B'}_C.$$

The analog to the Thomas' tractors comes from the standard representation of G on $\mathbb{T} = \mathbb{R}^{p+q} = \mathcal{V} \oplus \mathcal{W}$. Thus, fixing a Weyl connection, we get the tractors as couples $(v^A, w^{A'})$ with the transformation rules

$$\hat{v}^A = v^A - \Upsilon^A_{B'} w^{B'}, \quad \hat{w}^{A'} = w^{A'}.$$

Notice the special case $p = q = 2$ which provides (the split real form of) the Penrose's spinor presentation of tangent bundle and the two-component four-dimensional twistors $\mathcal{T}$. Indeed, $\mathfrak{so}(6, \mathbb{C}) = \mathfrak{sl}(4, \mathbb{C})$ and $\mathfrak{so}(4, \mathbb{C})$ splits into sum of two $\mathfrak{sl}(2, \mathbb{C})$ components. Thus, up to the choice of the right real form, the almost Grassmannian geometries with $p = q = 2$ correspond to the four-dimensional conformal Riemannian geometries.

The twistor parallel transport (connection) is then given by the formula

$$(\nabla^{\mathcal{T}})^A_{A'} \begin{pmatrix} v^B \\ w^{B'} \end{pmatrix} = \begin{pmatrix} \nabla^A_{A'} v^B + \mathsf{P}^{AB}_{A'C'} w^{C'} \\ \nabla^A_{A'} w^{B'} + \delta^{B'}_{A'} v^A \end{pmatrix}$$

and we shall see that this is the right formula for the standard tractor connection for the almost Grassmannian geometries in all dimensions.

The reader can find many further explicit examples in the last two chapters of [11], including those with non-trivial gradings on TM.

2.4 Elements of Tractor Calculus

In order to show how simple and general the basic functorial constructions and objects are, we shall focus for a while on general Cartan geometries with Klein models $G \to G/H$ without any further assumptions. But we shall come back to the parabolic and, in particular, conformal geometries in the end of this lecture.

2.4.1 Natural Bundles and Tractors

Let us come back to the functorial constructions on homogeneous spaces $G \to G/H$ mentioned in the first lecture. As always, $\mathfrak{h} \subset \mathfrak{g}$ are the Lie algebras of H and G.

For any Klein geometry G/H, there is the category of the homogeneous vector bundles, where the objects are the associated bundles $\mathcal{V} = G \times_H \mathbb{V}$. All morphisms on G/H are the actions of elements of G and these are mapped to the obvious actions on $\mathcal{V}$. Further morphisms in this category are the linear mappings intertwining the actions of the elements of G.

Clearly, there is the functor from the category of H-modules mapping the modules $\mathbb{V}$ to the associated bundles $\mathcal{V} = G \times_H \mathbb{V}$, while any module homomorphism $\phi : \mathbb{V} \to \mathbb{W}$ provides the morphisms $[\![u, v]\!] \longmapsto [\![u, \phi(v)]\!]$ between these bundles.

The latter functorial construction extends obviously to the entire category $\mathcal{C}_{G/H}$ of all Cartan geometries modelled on G/H. The morphisms have to respect the Cartan connections ω on the principal fiber bundles.

In this setting, a natural bundle is a functor $\mathcal{V} \colon \mathcal{C}_{G/H} \to \mathcal{VB}$ valued in the category of vector bundles. The functor sends every Cartan geometry $(\mathcal{G} \to M, \omega)$ to the vector bundle $\mathcal{V}M \to M$ over the same base (so it is a special case of the so-called gauge-natural bundles, see [17]). Moreover, $\mathcal{V}$ has the property that whenever there is a morphism between objects of $\mathcal{C}_{G/H}$, $\Phi \colon (\mathcal{G} \to M, \omega) \to (\tilde{\mathcal{G}} \to \tilde{M}, \tilde{\omega})$ covering $f \colon M \to \tilde{M}$, then there is the corresponding vector bundle morphism $\mathcal{V}\Phi \colon \mathcal{V}M \to \mathcal{V}\tilde{M}$ covering f. This is just an explicit description of the functoriality property with respect to the category of Cartan geometries. The main point is that each representation of H produces such a functor for all general Cartan geometries of the given type G/H.

At the same time, the Maurer–Cartan equation $d\omega + \frac{1}{2}[\omega, \omega] = 0$, valid on the homogeneous model, is no more true in general and we obtain the definition of the *curvature* κ of the Cartan geometries $(\mathcal{G} \to M, \omega)$ instead:

$$\kappa = d\omega + \frac{1}{2}[\omega, \omega]. \tag{2.37}$$

The fundamental Theorem 2.1 immediately reveals that a general Cartan geometry is locally isomorphic to its homogeneous model, if and only if its curvature vanishes identically.

We should also notice that there is the projective component of the curvature in $\mathfrak{g}/\mathfrak{h}$ which we call the *torsion*. Thus, the Cartan geometry is *torsion free* if the values of its curvature κ are in $\mathfrak{h}$. We shall see later that the normalizations of Cartan geometries consist in prescribing more complicated curvature restrictions, which always depend on the algebraic features of the Klein models.

As already mentioned, we are interested in specific functors on Cartan geometries $(\mathcal{G}, \omega)$ of the form $\mathcal{G} \times_H -$, referring to the associated bundle construction given for each fixed representation of H. See [11, section 1.5.5] for a detailed discussion on the topic of natural bundles on Cartan geometries. Specializing to representations of H which come as restrictions of representations of the whole group G leads to the following definition of *tractor bundles* below.

Recall the sections v of natural bundles $\mathcal{V}$ are identified with equivariant functions $v : \mathcal{G} \to \mathbb{V}$, i.e., $v(u \cdot g) = g^{-1} \cdot v(u)$. In particular, consider $\mathbb{V} = \mathfrak{g}/\mathfrak{h} = \mathbb{R}^n$ with the truncated adjoint action of H (i.e., the induced action on the quotient). The Cartan connection ω allows us to identify every tangent vector $\xi \in T_x M$ with the equivariant function $v : \mathcal{G} \to \mathbb{V}, u \mapsto \omega(\tilde{\xi}(u))$ for an arbitrary lift $\tilde{\xi}$ of ξ. This is the identification of the tangent bundle $TM \simeq \mathcal{V}M$. (And it completely justifies our earlier quite sloppy usage of elements in $\mathfrak{g}_{-1}$ instead of tangent vectors, etc.)

So in this way, the Cartan connection provides soldering of the tangent bundle, i.e., each element $u \in \mathcal{G}$ in the fiber over $x \in M$ can be viewed as a frame of $T_x M$. In general, different elements u may represent the same frame, depending on whether the truncated adjoint action of H on $\mathfrak{g}/\mathfrak{h}$ has got a non-trivial kernel.

Definition 2.2 The tractor bundles are natural vector bundles associated with the Cartan geometry $(\mathcal{G} \to M, \omega)$ of type $G \to G/H$, via restrictions of a representations of G to the subgroup H.

The unique principal connection form $\tilde{\omega} \in \Omega^1(\tilde{\mathcal{G}}) \to \mathfrak{g}$ on the extended principal G-bundle $\tilde{\mathcal{G}} = \mathcal{G} \times_H G$ extending the Cartan connection ω on $\mathcal{G}$ induces the so-called *tractor connections* $\nabla^{\mathcal{V}}$ on all tractor bundles $\mathcal{V}M$.

Notice that $\tilde{\mathcal{G}}$ is indeed a G-principal fiber bundle with the action of G defined by the right multiplication on the standard fiber G. Moreover, $u \mapsto [\![u, e]\!]$ provides the canonical inclusion of the principal fiber bundles $\mathcal{G} \subset \tilde{\mathcal{G}}$. The requested invariance of $\tilde{\omega}$, together with the reproduction of the fundamental vector fields, define the values of $\tilde{\omega}$ completely from its restriction $\tilde{\omega} = \omega$ on $T\mathcal{G}$.

In fact, we can equivalently define the tractor connections on the tractor bundles directly (by specifying their special properties), instead of referring to the Cartan connections on $\mathcal{G}$. This was also the approach by Thomas in [22]. The equivalence of such approaches for |1|-graded parabolic geometries was noticed and exploited in [14]. In full generality, the construction, normalization, and properties of tractor connections were derived in [8] (see also [11, Sections 1.5 and 3.1.22]).

2.4.2 Adjoint Tractors

A prominent example of tractor bundles arises when considering the Ad representation of the Lie group G on its Lie algebra $\mathfrak{g}$ and restricting it to H. Applying the corresponding associated bundle construction $\mathcal{G} \times_H -$ on the following short exact sequence of Lie algebras (with the obvious Ad actions)

$$0 \to \mathfrak{h} \to \mathfrak{g} \to \mathfrak{g}/\mathfrak{h} \to 0$$

we obtain

$$0 \to \mathcal{G} \times_H \mathfrak{h} \to \mathcal{AM} \overset{\pi}{\to} TM \to 0, \tag{2.38}$$

where we have identified $TM \cong \mathcal{G} \times_H \mathfrak{g}/\mathfrak{h}$. The middle term $\mathcal{AM} := \mathcal{G} \times_H \mathfrak{g}$ is called the *adjoint tractor bundle*.

Let us come back to the curvature (2.37) of the Cartan geometry now. Clearly we may evaluate κ on the so-called *constant vector fields* $\omega^{-1}(X)$ for all $X \in \mathfrak{g}$. Consider $X \in \mathfrak{h}$ and any $Y \in \mathfrak{g}$. Then $\omega^{-1}(X)$ is the fundamental vector field ζ_X and $d\omega(\zeta_X, -) = i_{\zeta_X} d\omega = \mathcal{L}_{\zeta_X}\omega = -\operatorname{ad}(X) \circ \omega$, by the equivariancy of ω. Thus,

$$\kappa(\omega^{-1}(X), \omega^{-1}(Y)) = \kappa(\omega^{-1}(X), \omega^{-1}(Y)) = -\operatorname{ad}(X)(Y) + [X, Y] = 0.$$

We have concluded that, actually, the curvature is a horizontal 2-form which can be represented by the equivariant *curvature function*

$$\kappa : \mathcal{G} \to \Lambda^2(\mathfrak{g}/\mathfrak{h})^* \otimes \mathfrak{g},$$
$$\kappa(X, Y)(u) = -\omega([\omega^{-1}(X), \omega^{-1}(Y)](u)) + [X, Y]. \tag{2.39}$$

In particular, we understand that the curvature descends to a genuine 2-form on the base manifold M valued in the adjoint tractors, i.e., $\kappa \in \Omega^2(M, \mathcal{AM})$.

There is much more to say about the adjoint tractors, we shall summarize several observations in the following two theorems (both were derived in [8], see also [11]).

Theorem 2.5

1. There is the (algebraic) Lie bracket $\{ , \}: \mathcal{AM} \times \mathcal{AM} \to \mathcal{AM}$ inherited from the Lie bracket on $\mathfrak{g}$.
2. The adjoint tractors are in bijective correspondence with the right-equivariant vector fields in $\mathcal{X}(\mathcal{G})^H$, and the Lie bracket of vector fields on $\mathcal{G}$ equips $\mathcal{AM}$ with the differential Lie bracket $[,]$, which is compatible with the Lie bracket on the tangent bundle TM, i.e., $\pi[\zeta, \eta] = [\pi\zeta, \pi\eta]$.[3]
3. If V is a tractor bundle, then there is the natural map $\bullet: \mathcal{AM} \times \mathcal{VM} \to \mathcal{VM}$, corresponding to the action of $\mathfrak{g}$ given by the G-representation $\mathbb{V}$. Moreover, $\{s_1, s_2\} \bullet t = s_1 \bullet s_2 \bullet t - s_2 \bullet s_1 \bullet t$.
4. The bracket $\{ , \}$ and the actions $\bullet$ are parallel with respect to the tractor connections $\nabla^{\mathcal{A}}, \nabla^{\mathcal{V}}$, i.e., for $s \in \mathcal{AM}$ and $v \in \mathcal{VM}$ we know

---

[3]Recall that $\pi : \mathcal{AM} \to TM$ is the projection from sequence (2.38).

$$\nabla_\xi^{\mathcal{A}}\{s_1, s_2\} = \{\nabla_\xi^{\mathcal{A}} s_1, s_2\} + \{s_1, \nabla_\xi^{\mathcal{A}} s_2\},$$

$$\nabla_\xi^{\mathcal{V}}(s \bullet v) = (\nabla_\xi^{\mathcal{A}} s) \bullet v + s \bullet (\nabla_\xi^{\mathcal{V}} v).$$

5. *For every tractor bundle* $\mathcal{V}$, *the value of the curvature* $R^{\mathcal{V}}$ *of the tractor connection* $\nabla^{\mathcal{V}}$ *is (for all vector fields* ξ, η *on* M *and sections* v *of* $\mathcal{V}M$)

$$R^{\mathcal{V}}(\xi, \eta)(v) = \kappa(\xi, \eta) \bullet v,$$

where $\kappa \in \Omega^2(M, \mathcal{A}M)$ *is the curvature of the Cartan connection.*

Proof The first claim is obvious just by definition. The Lie bracket on the Lie algebra is Ad-equivariant.

The adjoint tractors are smooth equivariant functions $\mathcal{G} \to \mathfrak{g}$. At the same time ω makes $T\mathcal{G}$ trivial. Now all $\xi \in T\mathcal{G}$ correspond to $\omega \circ \xi : \mathcal{G} \to \mathfrak{g}$ and the right-invariant fields ξ correspond just to the adjoint tractors. Since the Lie brackets of related fields are again related (here with respect to the principal actions of the elements in H), the Lie bracket restricts to $\mathcal{X}(G)^H$. Moreover, the right-invariant fields are projectable onto vector fields on M, and the same argument applies to brackets of the projections.

The third claim also follows directly from the definitions. Indeed, writing λ for the representation $\lambda : H \to \mathrm{GL}(\mathbb{V})$, and λ' for its differential at the unit, we recall $\exp(t \, \mathrm{Ad}(g)(X)) = g \exp(tX) g^{-1}$ and thus, differentiating we arrive at

$$\lambda'(\mathrm{Ad}(g)(X))(\lambda(g)(v)) = \lambda(g)(\lambda'(X)(v)).$$

Consequently, the bilinear map $\mathfrak{g} \times \mathbb{V} \to \mathbb{V}$ defined by λ' is G equivariant, it induces the map $\bullet : \mathcal{A}M \times \mathcal{V}M \to \mathcal{V}M$ and the bracket formula is just the defining property of a Lie algebra representation, in this picture.

The next claim is a straightforward consequence of the fact that both $\{\,,\,\}$ and $\bullet$ are operations induced by G-equivariant maps. Thus we may view them as living on the associated bundles to the extended G-principal fiber bundle $\tilde{\mathcal{G}}$. The formulae are just simple properties of the induced linear connections associated with a principal connection.

The same argument holds true for the last claim as well. $\square$

Notice also the definition of the operation $\bullet$ extends to all natural bundles $\mathcal{V}$, if we restrict the tractors only to the natural subbundle $\ker \pi \subset \mathcal{A}M$ of all vertical right invariant vector fields on $\mathcal{G}$, including the bracket compatibility property.

2.4.3 Fundamental Derivative

Consider the natural bundle $\mathcal{V} := \mathcal{G} \times_H \mathbb{V}$ associated with an H-representation λ on $\mathbb{V}$. Then, viewing the adjoint tractors as right-invariant vector fields on $\mathcal{G}$, we can define the differential operator $\mathrm{D} \colon \mathcal{A}M \times \mathcal{V}M \to \mathcal{V}M$ by the formula

$$\mathrm{D}_s\, v = s \cdot v,$$

where $s \in \mathcal{A}M$ is any tractor in $\mathcal{X}(\mathcal{G})^H$ differentiating the function $v \colon \mathcal{G} \to \mathbb{V}$. A simple check,

$$s(u \cdot h) \cdot v = (Tr^h \cdot s(u)) \cdot v = s(u) \cdot (v \circ r^h) = s(u) \cdot (\lambda_{h^{-1}} \cdot v) = \lambda_{h^{-1}}(s(u) \cdot v),$$

reveals that the result is again a smooth $\mathbb{V}$-valued H-equivariant mapping on $\mathcal{G}$. We call this operator D the *fundamental derivative*.

Notice that extending the tangent bundle to the adjoint tractors, we always have a canonical way of "differentiating" on all natural bundles for all Cartan geometries. As we may expect, there will be a lot of redundancy in such differentiation, since the vertical tractors in the kernel of the projection $\mathcal{A}M \to TM$ must act in an algebraic way due to the equivariance of the functions v.

Let us summarize some simple but very useful consequences of our definitions:

Theorem 2.6

1. *The fundamental derivative on the smooth functions (i.e., we consider the trivial representation $\mathbb{V} = \mathbb{R}$) is just the derivative in the direction of the projection:*

$$\mathrm{D}_s f = \pi(s) \cdot f.$$

2. *If the adjoint tractor s is vertical, i.e., $\pi(s) = 0$, then for every section v of a natural bundle $\mathcal{V}M$,*

$$\mathrm{D}_s\, v = -s \bullet v.$$

3. *The fundamental derivative D is compatible with all natural operations on natural bundles (i.e., those coming from H-invariant maps between the corresponding representation spaces). For example, having sections v, v^*, and w of natural bundles $\mathcal{V}$, $\mathcal{V}^*$, $\mathcal{W}$, and a function f*

$$\mathrm{D}_s(fv) = (\pi(s) \cdot f)v + f\,\mathrm{D}_s\, v$$

$$\mathrm{D}_s(v \otimes w) = \mathrm{D}_s\, v \otimes w + v \otimes \mathrm{D}_s\, w$$

$$\pi(s) \cdot v^*(v) = (\mathrm{D}_s\, v^*)(v) + v^*(\mathrm{D}_s\, v).$$

4. *If $\mathbb{V}$ is a G-representation, i.e., $\mathcal{V}$ is a tractor bundle, then*

$$\nabla^{\mathcal{V}}_{\pi(s)} v = \mathrm{D}_s\, v + s \bullet v.$$

Proof The equivariant functions $\mathcal{G} \to \mathbb{R}$ are just the compositions of functions f on the base manifold M with the projection $p : \mathcal{G} \to M$. Thus the first property is obvious, $s \cdot (f \circ p) = (Tp \cdot s) \cdot f = \pi(s) \cdot f$.

If s is vertical, then $s(u) = \zeta_Z(u)$, where ζ_Z is a fundamental vector field given by $Z \in \mathfrak{h}$. Thus,

$$s(u) \cdot v = \frac{d}{dt}\Big|_0 r^{\exp tZ}(u) \cdot v = -\lambda'(Z)(v(u)) = (-s \bullet v)(u).$$

The third property is again obvious—as long as the natural operations come from (multi)linear H-invariant maps, these will be compatible with the differentiations of functions valued in those spaces, in the directions of the right-invariant vector fields.

In order to see the last formula, consider a vector $\xi \in T_u\mathcal{G} \subset T_u\tilde{\mathcal{G}}$, covering a vector $\tau \in T_xM$. Then the horizontal lift of τ at the frame $u \in \mathcal{G} \subset \tilde{\mathcal{G}}$ is $\xi - \zeta_{\tilde{\omega}(\xi)} = \xi - \zeta_{\omega(\xi)}$. But the tractor connection is defined as the derivative of the equivariant function v in any frame of $\tilde{\mathcal{G}}$ in the direction of the horizontal lift and we obtain exactly the requested formula interpreting ξ as the value of the right-invariant vector field s (i.e., the adjoint tractor viewed as the equivariant function at u is expressed just via $\omega(\xi)$). $\qquad\qquad\qquad\qquad\qquad\qquad\qquad\qquad\qquad\qquad\qquad\qquad\quad\Box$

If we leave the slot for the adjoint tractor in the fundamental derivative free, we obtain the operator $\mathrm{D} : \mathcal{V}M \to \mathcal{A}^*M \otimes \mathcal{V}M$, and this can be obviously iterated,

$$\mathrm{D}^k : \mathcal{V}M \to \otimes^k \mathcal{A}^*M \otimes \mathcal{V}M.$$

Of course, there is a lot of redundancy in these higher order operators compared to standard jet spaces of the sections. In the case of the first order, we can identify the first jet prolongations $J^1\mathcal{V}$ of natural bundles as the natural bundles associated with the representations $J^1\mathbb{V}$ which are much smaller H-submodules in the modules $\mathbb{V} \oplus \mathrm{Hom}(\mathfrak{g}, \mathbb{V})$ corresponding to the values of the fundamental derivative. This is a useful observation because it implies that all invariant first order differential operators on the homogeneous models extend naturally to the entire category of the Cartan geometries with this model.

Before returning to the parabolic special cases, let us remark two more facts. The proofs are using similar arguments as above and the reader can find them in [11, sections 1.5.8, 1.5.9].

Expanding the formula for the exterior differential in the defining equation of the curvature κ, we can express the differential bracket on $\mathcal{A}M$:

$$
\begin{aligned}
[s_1, s_2] &= \mathrm{D}_{s_1} s_2 - \mathrm{D}_{s_2} s_1 - \kappa(\pi(s_1), \pi(s_2)) + \{s_1, s_2\} \\
&= \nabla^{\mathcal{A}}_{\pi(s_1)} s_2 - \nabla^{\mathcal{A}}_{\pi(s_2)} s_1 - \kappa(\pi(s_1), \pi(s_2)) - \{s_1, s_2\}.
\end{aligned}
\tag{2.40}
$$

There is the generalization of the well known Bianchi identities for curvature in the general Cartan geometry setting:

$$\sum_{\text{cyclic}} \left(\nabla^A_{\xi_1}(\kappa(\xi_2, \xi_3)) - \kappa([\xi_1, \xi_2], \xi_3) \right) = 0 \tag{2.41}$$

for all vector fields ξ_1, ξ_2, ξ_3, or its equivalent form for triples of adjoint tractors:

$$\sum_{\text{cyclic}} \left(\{s_1, \kappa(s_2, s_3)\} - \kappa(\{s_1, s_2\}, s_3) + \kappa(\kappa(s_1, s_2), s_3) + (D_{s_1} \kappa)(s_2, s_3) \right) = 0. \tag{2.42}$$

Similarly, the Ricci identity has got the general form for every section v of a natural bundle $\mathcal{V}$:

$$(D^2 v)(s_1, s_2) - (D^2 v)(s_2, s_1) = -D_{\kappa(s_1, s_2)} v + D_{\{s_1, s_2\}} v. \tag{2.43}$$

Notice, how easy we can read the classical identities for the affine connections from the latter two. Since the Cartan geometry is modelled on $\mathbb{R}^n = \text{Aff}(n, \mathbb{R})/\text{GL}(n, \mathbb{R})$ and the Lie algebra decomposes into direct sum of $\mathfrak{gl}(n, \mathbb{R})$-modules $\mathfrak{g}_{-1} = \mathbb{R}^n$ and $\mathfrak{g}_0 = \mathfrak{gl}(n, \mathbb{R})$, all the formulae decompose by homogeneities, $AM = TM \oplus P^1M$ (here P^1M is the linear frame bundle of TM), the bracket $\{\,,\,\}$ becomes trivial on TM, while the mixed bracket is just the evaluation. Thus, the Bianchi identity can be evaluated on tangent vectors and it decomposes into the two classical Bianchi identities for the torsion free connections, while it gets the more complex quadratic form in general. Similarly for Ricci, evaluated on s_1 and s_2 in TM. If the torsion is zero, κ has got only vertical values and thus the first term on the right-hand side is the algebraic action of the curvature (with plus sign), while the other one vanishes.

2.4.4 Back to Parabolic Geometries

Recall the parabolic cases always come with the splitting

$$\mathfrak{g} = \mathfrak{g}_- \oplus \mathfrak{p},$$

where $\mathfrak{g}_-$ is a subalgebra (but only a $\mathfrak{g}_0$ submodule). As before, we shall restrict ourselves to the $|1|$-graded case, although the below formulae easily extend to the general case.

Consider the category of parabolic geometries with the model G/P and a P-representation $\mathbb{V}$ which decomposes with respect to the action of the grading element $E \in \mathfrak{g}_0$ into $\mathbb{V} = \mathbb{V}_0 \oplus \cdots \oplus \mathbb{V}_k$. The adjoint tractor bundle has got the composition series

$$\mathcal{A}M = TM \oplus \operatorname{End} TM \oplus T^*M,$$

where the middle term is a subbundle in $T^*M \otimes TM$ corresponding to the group $G_0 = P/P_+$. Again, T^*M is the injecting part while TM is the projecting part, and the algebraic bracket $\{\ ,\ \}$ maps $T^*M \times TM \to \operatorname{End} TM$.

Once we fix a Weyl connection ∇, the Rho-tensor becomes a one-form valued in $T^*M \subset \mathcal{A}M$, we get the Rho-corrected derivative ∇^P, all P-modules get split into G_0-irreducible components which can be grouped according to the actions of the grading element in $\mathfrak{g}_0$, etc.

Theorem 2.7 *The fundamental derivative* D *on* $\mathcal{V}$ *is given in terms of any choice of Weyl connection by*

$$(\mathrm{D}_s v)_i = (\nabla^P_{\pi(s)} v)_i - s_0 \bullet v_i - s_1 \bullet v_{i-1} = \nabla_{\pi(s)} v_i - s_0 \bullet v_i + (P(\pi(s)) - s_1) \bullet v_{i-1},$$

where $s = (\pi(s), s_0, s_1)$ *and we indicate the splitting* $\mathcal{V} = \mathbb{V}_0 \oplus \cdots \oplus \mathbb{V}_k$ *with respect to the action of the grading element by the extra lower indices.*

If $\mathcal{V}$ *is a tractor bundle, then the tractor connection is given by*

$$(\nabla^{\mathcal{V}}_\xi v)_i = (\nabla^P_\xi v)_i + \xi \bullet v_{i+1} = \nabla_\xi v_i + P(\xi) \bullet v_{i-1} + \xi \bullet v_{i+1}.$$

Proof Both formulae are direct consequences of the general formulae and the definitions. The reader may also consult [11, section 5.1.10]. $\qquad\square$

2.4.5 Towards Effective Calculus for Conformal Geometry

Now, with the general concepts and formulae at hand, it is obvious that the Thomas' tractors come equipped with the nice tractor connection on all conformal Riemannian manifolds in the sense of Cartan geometries and the connection will be always given by the formulae in Theorem 2.2, which are manifestly invariant. Moreover, we know that the curvature of the Thomas' tractor connection on the sphere (with the Maurer–Cartan form ω) is zero.

But we still cannot be happy enough, for at least two reasons. First, we want to define the geometries by a structure on the tangent bundle and we shall come to that question in the next lecture. Second, we need some more effective manifestly natural operators than the fundamental derivative.

We shall only briefly comment on the latter problem and advise the readers to look at [12] for much more information.

Already Tracy Thomas constructed the differential operator D which is invariant for $\sigma \in \mathcal{E}[1]$, with values in $\mathcal{T}$ (we follow the usual convention of [12] and write the projecting part in the top, while the injecting part is in the bottom of the column vector). We may follow our prolongation of the "conformal to Einstein" equation from the second lecture. Starting with σ in $\mathcal{E}[1]$, we first put $\mu_a = \nabla_a \sigma$ in $\mathcal{E}_a[1]$ and then, contracting the equation $\nabla_a \nabla_b \sigma + P_{ab}\sigma + g_{ab}\rho = 0$ we see $-n\rho = \nabla^a \nabla_a \sigma + P^a{}_a \sigma$. Thus, adjusting the $1/n$ factor, we arrive at the operator $D : \mathcal{E}[1] \to \mathcal{T}$

$$\sigma \xrightarrow{D} \begin{pmatrix} n\sigma \\ n\nabla_a\sigma \\ -(\nabla^a\nabla_a + P^a{}_a)\sigma \end{pmatrix}. \tag{2.44}$$

This *Thomas' D-operator* extends to all densities $\mathcal{E}[w]$. For $f \in \mathcal{E}[w]$ we define Df in $\mathcal{T}[w-1]$ as

$$Df = \begin{pmatrix} (n+2w-2)wf \\ (n+2w-2)\nabla_a f \\ -(\nabla^a\nabla_a + wP^a{}_a)f \end{pmatrix}. \tag{2.45}$$

In particular, we should notice the following facts. For $w = 0$, the first non-zero slot in the column is $(n-2)\nabla_a f$. Thus, this operator must be invariant and we have recovered the usual differential of functions.

A much more interesting choice is $w = 1 - \frac{n}{2}$ since this kills the first two components and the third one gets manifestly invariant. This way we get the second order operator $\nabla^a\nabla_a + \frac{2-n}{2}P^a{}_a$ and we recognize the celebrated Yamabe operator mentioned already in the third lecture. (Just checking the pedestrian way the invariance of this operator shows that the general theory was worth the effort!)

This example indicates where the genuine tractor calculus goes with the aim to construct manifestly invariant operators in an effective way.

2.5 The (Co)homology and Normalization

We shall continue with parabolic $P \subset G$ and the Klein model $G \to G/P$, mainly restricting to $|1|$-graded $\mathfrak{g}$. Thus $\mathfrak{g} = \mathfrak{g}_- \oplus \mathfrak{g}_0 \oplus \mathfrak{p}_+ = \mathfrak{g}_{-1} \oplus \mathfrak{g}_0 \oplus \mathfrak{g}_1$.

Recall that any choice of the reduction $\sigma : \mathcal{G}_0 = \mathcal{G}/\exp\mathfrak{g}_1 \to \mathcal{G}$ of the structure group of a Cartan geometry $(\mathcal{G} \to M, \omega)$ provides the pullback $\sigma^*(\omega)$ which splits into the soldering form $\theta \in \Omega^1(\mathcal{G}_0, \mathfrak{g}_{-1})$ (independent of the choice of σ), the Weyl connection ∇_a, and the Rho-tensor P_{ab}, which is a T^*M valued one-form on M. Moreover, the adjoint tractor bundle splits as

$$\mathcal{A}M = TM \oplus \mathcal{A}_0 M \oplus T^*M.$$

Our aim is now to find some suitable normalization allowing to construct a natural Cartan connection from the data on $\mathcal{G}_0$. Once we succeed, the tractor calculus related to this Cartan connection will become a natural part of the geometry defined on $\mathcal{G}_0$. We shall see that the crucial tool at our disposal is related to the cohomological properties of the Lie algebras in question. There are two equivalent ways: either to normalize the curvature of the Cartan connection or to normalize the curvature of a suitable tractor connection. We shall show the first one, the other one was first achieved in [8], and both are explained in full generality in [11, chapter 3].

2.5.1 Deformations of Cartan Connections

The obvious idea is to quest for normalizations which will make the curvatures of the Cartan connections as small as possible. In particular, this will ensure that the right Cartan connections on homogeneous models will be the Maurer–Cartan forms.

Consider two Cartan connections on the same principal bundle $\mathcal{G} \to M$, ω and $\tilde{\omega}$. Then their difference $\Phi = \tilde{\omega} - \omega$ clearly vanishes on all vertical vectors and is right-invariant. Thus, we deal with a one-form $\Phi \in \Omega^1(M, \mathcal{A}M)$.

In the $|1|$-graded case, let us understand the "geometry" on M as the choice of the G_0-principal bundle $\mathcal{G}_0$ together with the soldering form θ, i.e., we adopt the most classical concept of a G-structure as a reduction of the first order linear frame bundle P^1M to the structure group G_0. (We already mentioned in the examples in lecture 3 that the projective geometries are different.) It is obvious from our definitions that the two Cartan connections will define the same structure in the latter sense if and only if their difference has got values in $\mathfrak{p}$. Thus, in our $|1|$-graded cases, Φ should be in $\Omega^1(M, \mathcal{A}_0M \oplus T^*M)$.

In the general situation with longer gradings, we have to be much more careful with the definition of the G_0-structure which has to be generalized to the filtered manifolds. In brief, the tangent space inherits the filtration by $\mathfrak{p}$-submodules of $\mathfrak{g}_-$ and a full analog of the classical G-structure has to be considered on the associated graded vector bundle $\operatorname{Gr} TM$. We shall not go to any details here, the reader can find a detailed exposition in [11, chapter 3].

As we know, the curvature can be also viewed as the curvature function $\kappa : \mathcal{G} \to \Lambda^2(\mathfrak{g}/\mathfrak{p})^* \otimes \mathfrak{g}$, and $(\mathfrak{g}/\mathfrak{p})^* = \mathfrak{p}_+$ via the Killing form on $\mathfrak{g}$. Thus, we should like to know how κ changes if we deform the Cartan connection by Φ in $\Omega^1(M, \mathcal{A}_0M \oplus T^*M)$.

Let us write κ_ℓ for the component of the curvature function of homogeneity ℓ, i.e., $\kappa_\ell \in \Lambda^2\mathfrak{g}^*_{-1} \otimes \mathfrak{g}_{\ell-2}$ for the $|1|$-graded parabolic geometries.

Lemma 2.2 *Assume $\Phi \in \Omega^1(M, \mathcal{A}_0M \oplus T^*M)$ is of homogeneity $\ell = 1$ or $\ell = 2$. Then the components of the curvature of homogeneities lower than ℓ remain unchanged, while the corresponding change of the $\mathfrak{g}_{-1}$ or $\mathfrak{g}_0$ component of the curvature, viewed as function valued in $\Lambda^2\mathfrak{g}^*_{-1} \otimes \mathfrak{g}_i$ with $i = -1$ or 0, respectively, is given by the formula*

$$(\tilde{\kappa} - \kappa)_i(X, Y) = [X, \phi(Y)] - [Y, \phi(X)],$$

where ϕ is the equivariant function $\mathcal{G} \to \mathfrak{g}^*_{-1} \otimes (\mathfrak{g}_0 \oplus \mathfrak{g}_1)$ representing Φ.

Proof Considering vector fields $\xi, \eta \in T\mathcal{G}$,

$$\tilde{\omega}(\xi) = \omega(\xi) + \phi(\omega(\xi)).$$

Thus, hitting the equation with the exterior derivative, we obtain

$$d\tilde{\omega}(\xi, \eta) = d\omega(\xi, \eta) + d\phi(\xi)(\omega(\eta)) - d\phi(\eta)(\omega(\xi)) + \phi(d\omega(\xi, \eta)),$$

while

$$[\tilde{\omega}(\xi), \tilde{\omega}(\eta)] = [\omega(\xi), \omega(\eta)] + [\phi(\omega(\xi)), \omega(\eta)] + [\omega(\xi), \phi(\omega(\eta))] + [\phi(\omega(\xi)), \phi(\omega(\eta))].$$

Comparing the curvatures (as $\mathfrak{g}$-valued two forms on $\mathcal{G}$),

$$(\tilde{\kappa} - \kappa)(\xi, \eta) = d\phi(\xi)(\omega(\eta)) - d\phi(\omega(\eta))(\xi) + \phi(d\omega(\xi, \eta))$$
$$- [\phi(\omega(\xi)), \omega(\eta)] + [\omega(\xi), \phi(\omega(\eta))] + [\phi(\omega(\xi)), \phi(\omega(\eta))].$$

Now, inspecting the homogeneities for ϕ valued in $\mathfrak{g}_i$ ($i = 0$ corresponds to homogeneity 1, while $i = 1$ yields homogeneity 2), the first three terms will land in $\mathfrak{g}_i$, while the very last term is either zero (if $i = 1$) or sits in $\mathfrak{g}_i$ again (if $i = 0$). Thus only the two remaining brackets have got the values in $\mathfrak{g}_{i-1}$ and we obtain just the requested result if we write the vector fields as functions on $\mathcal{G}$ with the help of ω.   □

2.5.2   Homology and Cohomology

The formula for the lowest homogeneity deformation of the curvature is a special instance of a general algebraic construction, which works for arbitrary Lie algebra $\mathfrak{g}$ and $\mathfrak{g}$-module $\mathbb{V}$. We define the k-chains $C_k(\mathfrak{g}, \mathbb{V})$ as

$$C_k(\mathfrak{g}, \mathbb{V}) := \Lambda^k \mathfrak{g} \otimes \mathbb{V}.$$

For each $k > 0$ we define the linear operator $\delta_k \colon C_k \to C_{k-1}$

$$\delta_k(X_1 \wedge \cdots \wedge X_k \otimes v) = \sum_i (-1)^i \underbrace{X_1 \wedge \cdots \wedge X_k}_{\text{omit } i\text{-th}} \otimes X_i \cdot v$$

$$+ \sum_{i<j} (-1)^{i+j} [X_i, X_j] \wedge \underbrace{X_1 \wedge \cdots \wedge X_k}_{\text{omit } i\text{-th, } j\text{-th}} \otimes v.$$

Then $\delta^2 = 0$ and thus δ acts on the chain complex $C(\mathfrak{g}, \mathbb{V})$ as a boundary operator. A direct check reveals that δ is always a $\mathfrak{g}$-module homomorphism. Therefore we can define the homology groups

$$H_k(\mathfrak{g}, \mathbb{V}) = \frac{\ker \delta_k}{\operatorname{im} \delta_{k+1}}$$

and they are again $\mathfrak{g}$-modules.

Note that $C_0(\mathfrak{g}, \mathbb{V}) = \mathbb{V}$ and $C_1(\mathfrak{g}, \mathbb{V}) \xrightarrow{\delta_1} C_0(\mathfrak{g}, \mathbb{V})$ is given by $\delta_1(X \otimes v) = X \cdot v$ which implies $H_0(\mathfrak{g}, \mathbb{V}) = \mathbb{V}/\langle X \cdot v \rangle = \mathbb{V}/\mathfrak{g} \cdot \mathbb{V}$.

In particular, considering the adjoint representation, $H_0(\mathfrak{g}, \mathfrak{g}) = \mathfrak{g}/[\mathfrak{g}, \mathfrak{g}]$.

Similarly to the homology, we can consider the dual construction for cochains $C^k(\mathfrak{g}, \mathbb{V}) = \Lambda^k \mathfrak{g}^* \otimes \mathbb{V}$ and coboundaries $\partial_k : C^k(\mathfrak{g}, \mathbb{V}) \to C^{k+1}(\mathfrak{g}, \mathbb{V})$ given by

$$\partial_k \varphi(X_0, \ldots, X_k \otimes v) = \sum_i (-1)^i X_i \varphi(\underbrace{X_0, \ldots, X_k}_{\text{omit } i\text{-th}}) \otimes X_i \cdot v$$

$$+ \sum_{i<j} (-1)^{i+j} \varphi([X_i, X_j], \underbrace{X_1, \ldots, X_k}_{\text{omit } i\text{-th and } j\text{-th}}) \otimes v \, .$$

Then ∂ provides a coboundary operator on the complex of cochains, i.e., $\partial^2 = 0$. The operators ∂ are again $\mathfrak{g}$-module homomorphisms and we define the cohomology groups

$$H^k(\mathfrak{g}, \mathbb{V}) = \frac{\ker \partial_k}{\operatorname{im} \partial_{k-1}}.$$

Again, the zero cohomology is easy to compute. Clearly $\partial_0(v)(X_0) = X_0 \cdot v$, while

$$\partial_1 \psi(X, Y) = X \cdot \psi(Y) - Y \cdot \psi(X) - \psi([X, Y]).$$

Thus, $H^0(\mathfrak{g}, \mathbb{V}) = \mathbb{V}^\mathfrak{g} \subset \mathbb{V}$ is the kernel of the $\mathfrak{g}$-action. If we choose $\mathbb{V} = \mathfrak{g}$ with the adjoint action, then $H^1(\mathfrak{g}, \mathfrak{g}) = \{\text{all derivatives}\}/\{\text{inner derivatives}\}$.

Now, the crucial observation is that Lemma 2.2 expresses the lowest homogeneity of the deformation of the curvature of our Cartan geometries, caused by $\phi \in \mathfrak{g}_{-1}^* \otimes \mathfrak{g}_i$, via the coboundary differential $\partial \phi$ (the third term is not there in our case since we deal with $|1|$-graded geometries).

For general parabolic geometries we also consider the curvature as an equivariant function $\kappa : \mathcal{G} \to C^2(\mathfrak{g}_- \otimes \mathfrak{g})$ and $\mathfrak{g}$ is a $\mathfrak{g}_-$-module with the adjoint action. Even in full generality, the Lemma 2.2 holds true, i.e., the lowest homogeneity of the curvature deformation caused by ϕ is given by $\partial \phi$, see [11, section 3.1.10].

2.5.3 Normalization of Parabolic Geometries

We should be interested in the cohomologies $H^k(\mathfrak{g}_-, \mathfrak{g})$, in particular in the second degree since the curvature has got the values in the second degree cochains. Recall Lemma 2.2 which discussed how all possible deformations of the Cartan curvature (with positive homogeneities) impact the curvature. In particular, we learned there that the available deformation of the curvature fills the image of ∂ in the second degree cochains (in the lowest non-trivial homogeneity).

Now the crucial moment comes. Consider parabolic geometries with the homogeneous model G/P, $\mathfrak{g} = \mathfrak{g}_- \oplus \mathfrak{g}_0 \oplus \mathfrak{p}_+$ and a $\mathfrak{g}$-module $\mathbb{V}$. Recall $(\mathfrak{g}/\mathfrak{p})^* \cong \mathfrak{g}_-^* \cong \mathfrak{p}_+$. Thus, the dual of the space of cochains $C^k(\mathfrak{g}_-, \mathbb{V})$ is $C^k(\mathfrak{p}_+, \mathbb{V}^*)$ and there is the dual mapping $\partial^* : C^{k+1}(\mathfrak{p}_+, \mathbb{V}^*) \to C^k(\mathfrak{p}_+, \mathbb{V}^*)$. It was Kostant who noticed in his celebrated paper [18] that there always is a scalar product $\langle\ ,\ \rangle$ on the space of cochains $C^k(\mathfrak{p}_+, \mathbb{V}^*)$ such that, identifying $C^k(\mathfrak{p}_+, \mathbb{V}^*)$ with $C^k(\mathfrak{g}_-, \mathbb{V})$, the latter dual map ∂^* becomes the adjoint operator to ∂. Moreover its formula then coincides with the boundary operator δ. We shall follow the (confusing) convention by many authors and call this adjoint ∂^* the *codifferential*. In particular, ∂^* is a P-module homomorphism.

Now, we equivalently consider

$$H^k(\mathfrak{g}_-, \mathbb{V}) = H^k(\mathfrak{p}_+, \mathbb{V}^*) = \frac{\ker \partial^*}{\operatorname{im} \partial^*}$$

and, applying the standard algebraic Hodge theory, we get the decompositions (of G_0-modules)

$$C^k(\mathfrak{g}_-, \mathbb{V}) = \operatorname{im} \partial^* \oplus \ker \partial = \ker \partial^* \oplus \operatorname{im} \partial = \operatorname{im} \partial^* \oplus \ker \square \oplus \operatorname{im} \partial, \qquad (2.46)$$

where $\square \equiv \partial\partial^* + \partial^*\partial$ (thus the intersection of the kernels of ∂ and ∂^*). This means that the cohomology $H^k(\mathfrak{p}_+, \mathbb{V}^*) = H^k(\mathfrak{g}_-, \mathbb{V})$ equals to the kernel of the algebraic Hodge Laplacian operator $\square$.

Further, we see that $\ker \partial^*$ is always the complementary subspace to $\operatorname{im} \partial$ and in view of Lemma 2.2 we adopt the following normalization.

Notice ∂^* is a P-module homomorphism and so it induces natural transformations between the corresponding natural bundles. In particular, it makes sense to apply ∂^* to the curvatures of our Cartan connections, i.e., there is the natural algebraic operator

$$\partial^* : \Lambda^2 T^*M \otimes \mathcal{A}M \to T^*M \otimes \mathcal{A}M$$

which preserves the homogeneities.

Definition 2.3 Let $(\mathcal{G} \to M, \omega)$ be a parabolic geometry with the homogeneous model $G \to G/P$, $\mathfrak{g} = \mathfrak{g}_{-k} \oplus \cdots \oplus \mathfrak{g}_k$. We say that ω is a *regular parabolic* *geometry*, if its curvature κ has got only positive homogeneities. The geometry is called *normal*, if its curvature is co-closed, i.e., $\partial^*\kappa = 0$.

Let us stress the following observation. The curvature of any normal parabolic geometry lies in the kernel of ∂^* and thus it projects to the natural bundle defined by the cohomology $H^2(\mathfrak{g}_-, \mathfrak{g})$. This is the so-called *harmonic curvature* $\kappa^H \in \mathcal{G} \times_P H^2(\mathfrak{g}_-, \mathfrak{g})$.

Let us restrict again our attention to $|1|$-graded geometries. First notice, the regularity condition is empty in this case. Indeed, the decomposition of κ into its homogeneity components coincides with the decomposition by its values, i.e., values in $\mathfrak{g}_i$ are of homogeneity $i + 2$, $i = -1, 0, 1$.

Further, there is a nice consequence of the Bianchi identity (2.42). Consider the component κ_i of the lowest homogeneity ℓ. Then the four terms in (2.42) are of homogeneity at least, $\ell - 1$, $\ell - 1$, ℓ, ℓ, respectively. But each homogeneity component in (2.42) has to vanish independently. Finally, the first two terms represent exactly the differential $\partial\kappa$.

We conclude that the lowest homogeneity non-zero component of the curvature should be closed and thus, for normal geometries it must coincide with its harmonic projection. Moreover, if all these harmonic components are zero, then we conclude (by induction using the previous result) that the entire curvature κ must vanish, too. These results hold true even for general parabolic geometries, the reader may consult [11, section 3.1.12].

Now we are ready to manage the normalization of the $|1|$-graded parabolic geometries with $\mathfrak{g} = \mathfrak{g}_{-1} \oplus \mathfrak{g}_0 \oplus \mathfrak{g}_1$. Given any G_0-principal bundle $\mathcal{G}_0 \to M$ with the soldering form $\theta \in \Omega^1(M, \mathfrak{g}_{-1})$, i.e., a classical G_0-structure, we consider the fiber bundle $\mathcal{G} = \mathcal{G}_0 \times \exp\mathfrak{g}_1$ and equip it with the obvious principal action of $P = G_0 \ltimes \exp\mathfrak{g}_1$.

If we choose any principal connection γ on $\mathcal{G}_0$, then $\theta \oplus \gamma$ is a Cartan connection on $\mathcal{G}_0 \subset \mathcal{G}$ and choosing any $\mathsf{P} \in \Omega^1(M, T^*M)$, there is exactly one Cartan connection ω on $\mathcal{G}$ coinciding with $\theta \oplus \gamma \oplus \mathsf{P}$ on $T\mathcal{G}_0 \subset T\mathcal{G}$.

The connection is automatically regular and the lowest component of its curvature can have homogeneity 1. It is a simple exercise to see that this component will coincide with the torsion T of the connection γ (e.g., viewed as the torsion part of the curvature of the Cartan connection $\theta \oplus \gamma$). Moreover, changing the inclusion of $\mathcal{G}_0 \to \mathcal{G}$, i.e., choosing a Weyl connection for ω, this torsion part does not change at all.

We know that for the normal Cartan connections, this torsion has to coincide with its harmonic part. Moreover, Lemma 2.2 says that we can modify the Cartan connection ω by a homogeneity one deformation Φ so that this condition will be satisfied.

In fact, this only recovers the very classical results about the distinguished connections with special torsions on G-structures.

For example, in conformal Riemannian geometry, there is no cohomology in homogeneity one and thus we may always find torsion free connections. This is, of course, no surprise since we may take any Levi-Civita connection of one of the metrics in the class. But for the almost Grassmannian geometries with $p \geq q \geq 3$, all the cohomology appears in homogeneity one only (with two irreducible components) and thus connections with torsions are unavoidable in general, unless we deal with the homogeneous models.

Next, we may assume that we have chosen the above connection γ in such a way that its torsion is harmonic. In order to see the link between the curvature of γ and the curvature κ of ω, consider the Cartan connection $\tilde{\omega}$ on $\mathcal{G}$ which would be given by the choice $\mathsf{P} = 0$. The Cartan connections $\theta \oplus \gamma$ and $\tilde{\omega}$ are related by the inclusion $\mathcal{G}_0 \to \mathcal{G}$ and thus the curvature $\tilde{\kappa}$, restricted to $\mathcal{G}_0$ coincides with the curvature $T + R$ of $\theta \oplus \gamma$. Thus, Lemma 2.2 says (with the deformation $\mathsf{P} = \omega - \tilde{\omega}$) that the homogeneity two component of the curvature of ω is

$$\kappa_0 = R + \partial \mathsf{P}.$$

Hitting this equality by ∂^* gives

$$\partial^* \kappa_0 = \partial^* R + \partial^* \partial \mathsf{P}.$$

But by homogeneity argument, $\partial^* \mathsf{P}$ would have values in $\mathfrak{g}_2$ and thus vanishes automatically. Thus, the second term in the latter equation equals $\square \mathsf{P}$ and the normalization condition will be satisfied if we choose P such that

$$\square \mathsf{P} = -\partial^* R. \tag{2.47}$$

The final crucial observation is that the Laplacian acts by non-zero constant multiples on all irreducible components, except the harmonic ones. But we want to invert $\square$ on im ∂^*, which cannot include any harmonic components. The final formula for P is

$$\mathsf{P} = -\square^{-1} \partial^* R. \tag{2.48}$$

Summarizing, in order the construct the normal Cartan connection ω on a manifold equipped with the relevant G_0-structure, we first choose any connection γ with harmonic torsion. Then we consider its curvature R, apply the codifferential and compute the right coefficients for each of its irreducible components. There are effective tools in the representation theory allowing to compute them easily via the so-called Casimir operators. We have no space to go into details here.

Finally, there is the question about the uniqueness of our construction. The answer is again hidden in cohomologies. If there are no positive homogeneity components in $H^1(\mathfrak{g}_{-1}, \mathfrak{g})$, all our choices of the deformations in both steps were unique. This is the case for nearly all $|1|$-graded geometries. The only exceptions are the projective geometries (and their complex versions), where we have to choose one of the connections in the first step to define the structure. Then the Cartan connection is already given uniquely via the next step in our construction.

In the categorical language, there is the subcategory of the regular and normal Cartan geometries, and this subcategory is equivalent to the category of the infinitesimal G_0-structures on manifolds, up to some rare exceptions due to the existence of positive homogeneities in first cohomologies in some examples (where a similar equivalence exists, too).

In conformal Riemannian geometry, i.e., $\mathfrak{g} = \mathfrak{so}(n + 1, 1)$, there is no positive homogeneity first cohomology, while the entire second cohomology is concentrated in homogeneity two (except of dimension $n = 3$, where it is homogeneity three). The operator ∂^* is just the trace, so the image on the curvature of a Levi-Civita connection is the Ricci tensor. The formula for P reflects the right choices of the constants in the action of $\square$, while the invariant Weyl part of the curvature (shared by all Weyl connections) is $R + \partial \mathsf{P}$, the harmonic component in all dimensions $n > 3$. Of course, the geometry is locally isomorphic to the conformal sphere if and only if this Weyl curvature vanishes.

We do not have space in this lecture to inspect further examples and detailed computations. The readers may look up many of them in [11], a few hundreds of pages of examples and details for general parabolic geometries are there in chapters 3 through 5.

2.6   The BGG Machinery

As well known, the linearized theories in Physics usually appear as locally exact complexes of differential operators. A lot of attention was devoted to this phenomenon in Mathematics, too. Already in the early days people around Gelfand or Kostant knew that on the Klein models, the existence of such complexes is an algebraic phenomenon related to homomorphisms of Verma modules (which were understood as topological duals of the infinite jet prolongations of the natural bundles), cf. [4, 19].

The main message of this series of lectures is to show how remarkably the algebraic features and phenomena from the Klein models extend to the categories of Cartan geometries. The so-called BGG machinery does exactly this—extends the complexes of the differential operators from the homogeneous models to sequences on all Cartan geometries of the given type.

In this last lecture we comment on this exciting development and we shall also come back to the solutions of the "conformal to Einstein" equation (2.8) in terms of constant tractors. On the way we shall touch the general construction of the latter sequences of operators and identify Eq. (2.8) as one of the so-called 1st BGG operators.

2.6.1 The Twisted de-Rham Complexes

Denote by $H_{\mathbb{V}}^k M$ the natural bundle associated with the P-module $H^k(\mathfrak{g}_-, \mathbb{V})$ of cohomologies with coefficients in a G-module $\mathbb{V}$. Notice that by the Kostant's complete description of the cohomologies [18], the latter cohomology module is a G_0 module with trivial action of P_+ and thus, it is completely reducible. In particular, $H_{\mathbb{V}}^0 M$ is the bundle coming from the projecting part of $\mathbb{V}$ which can be viewed as the orbit of the lowest weight vector in $\mathbb{V}$ under the $\mathfrak{g}_0$-action. Our goal is to come to the following diagram of operators

$$
\begin{array}{ccc}
\Omega^0(M, \mathcal{V}M) & \xrightarrow{\ \mathrm{d}_{\mathbb{V}}\ } & \Omega^1(M, \mathcal{V}M) \xrightarrow{\ \mathrm{d}_{\mathbb{V}}\ } \dots \\
\pi \downarrow \uparrow \mathrm{L} & & \pi \downarrow \uparrow \mathrm{L} \\
H_{\mathbb{V}}^0 M & \xdashrightarrow{\ D\ } & H_{\mathbb{V}}^1 M \xdashrightarrow{\ D\ } \dots
\end{array}
$$

$$\tag{2.49}$$

where all the arrows have to be yet explained. As usual, we write $\mathcal{V}M$ for the tractor bundle over the manifold M corresponding to $\mathbb{V}$, and notice that ∂^* is the adjoint of ∂ which is a P-module homomorphism and thus, it gives rise to the natural algebraic operator $\partial^* : \Omega^k(M, \mathcal{V}M) \to \Omega^{k-1}(M, \mathcal{V}M)$. Clearly, the projections π are well defined only on the kernel of ∂^*. We shall have to be careful about this.

The ideas presented below go back to [2] and [3], and they were further developed in [10].

Let us discuss the upper line in (2.49) now. First, restrict to the parabolic Klein model $G \to G/P$. Together with the G-module $\mathbb{V}$, consider a P-module $\mathbb{W}$. Then there is the following identification of the sections of the tensor product should appear in the opposite order: $\mathbb{W} \otimes \mathbb{V}$. For any section s of $\mathbb{W}$, i.e., an equivariant mapping $s : \mathcal{G} \to \mathbb{W}$, and $v \in \mathbb{V}$ consider the map

$$
s \otimes v \longmapsto \underbrace{(g \longmapsto s(g) \otimes g^{-1} \cdot v)}_{\text{equivariant } \mathcal{G} \to \mathbb{W} \otimes \mathbb{V}},
$$

which provides a natural isomorphism of the G-modules of sections

$$
\Gamma(\mathcal{W}) \otimes \mathbb{V} \cong \Gamma(\mathcal{W} \otimes \mathcal{V}). \tag{2.50}
$$

Thus, if $F : \mathcal{W}_1 \to \mathcal{W}_2$ is an arbitrary differential operator between the homogeneous vector bundles, then $F \otimes \mathrm{Id}_{\mathbb{V}} = F_{\mathbb{V}}$ provides the *twisted operator* $F_{\mathbb{V}} : \mathcal{W}_1 \otimes \mathcal{V} \to \mathcal{W}_2 \otimes \mathcal{V}$.

Considering the exterior differential $\mathrm{d} : \Lambda^k T^*M \to \Lambda^{k+1} T^*M$, this explains the whole first line in (2.49), at least on the homogeneous model. On zero-degree forms, the exterior differential is just the covariant derivative of the sections.

Let us look more carefully on this example. At the level of first order jets, we can express the twisted operator by means of the algebraic P-homomorphism

$$J^1(\Lambda^k \mathfrak{p}_+ \otimes \mathbb{V}) \to \Lambda^{k+1} \mathfrak{p}_+ \otimes \mathbb{V}, \quad (f_0, Z \otimes f_1) \longmapsto \partial f_0 + (k+1) Z \wedge f_1. \quad (2.51)$$

In general, if we write $J^r(\mathbb{W})$ and $\bar{J}^r(\mathbb{W})$ for the standard fibers of the holonomic and semi-holonomic jet prolongations $J^r(\mathcal{W})$, $\bar{J}^r(\mathcal{W})$,[4] then the isomorphism (2.50) must hold true at the jet level, e.g., $\bar{J}^r(\mathbb{W}) \otimes \mathbb{V} \cong \bar{J}(\mathbb{W} \otimes \mathbb{V})$.

Now the crucial observation comes: Although the jet prolongations $J^r \mathcal{W}$ are no more natural bundles associated with $\mathcal{G}$ in general, there is still no problem with the first jets. Thus, $J^1(\mathcal{W}) = \mathcal{G} \times_P J^1(\mathbb{W})$ and iterating this procedure, we conclude that the semi-holonomic jet prolongations are natural bundles again, i.e., $\bar{J}^r(\mathcal{W}M) = \mathcal{G} \times_P \bar{J}^r(\mathbb{W})$ for the relevant P-module $\bar{J}^r(\mathbb{W})$ (the standard fiber over the origin in G/P as the module with the action of the isotropy group P). Moreover, we can construct a universal differential operator $\mathcal{W}M \to \bar{J}^r(\mathcal{W}M)$ based on the iterated fundamental derivative, which allows one to extend many invariant operators from the homogeneous model to all Cartan geometries of this type.

Therefore, the so-called *strongly invariant operators*, i.e., those coming from algebraic P-module homomorphisms $\bar{J}^r(\mathbb{W}_1) \to \mathbb{W}_2$, enjoy a canonical extension to all Cartan geometries by means of the formulae obtained on the homogeneous model.

A careful exposition of the algebraic structure of the semi-holonomic jets and their links to the strongly invariant operators can be found in [13].

This in particular applies for all first order operators and we are done with the first line in (2.49), which is called the *twisted de-Rham* sequence. Obviously, there are many other ways for twisting the de-Rham. For example, we could take the covariant exterior differential d^ω of the tractor valued k-forms with respect to the tractor connection on $\mathcal{V}$. A straightforward computation reveals

$$d^\omega \varphi = d_{\mathbb{V}} \varphi + \iota_{\kappa_-} \varphi, \quad (2.52)$$

where κ_- is the torsion part of the curvature $\kappa = d\omega + \frac{1}{2}[\omega, \omega]$.

2.6.2 BGG Machinery

Next, let us focus on the vertical arrows in (2.49). We already know about the projections π, so we have to deal with L's.

---

[4]We iterate the first jet prolongation. Considering the first jets of sections of a bundle $\mathcal{W}$, the jets in a fiber of $J^1(J^1\mathcal{W})$ look in coordinates as 4-tuples $(y^p, y_i^p, Y_j^p, Y_{ij}^p)$ were Y_{ij}^p do not need to be symmetric. These are the non-holonomic 2-jets. The semi-holonomic ones remove part of the redundancy by requesting that the two natural projections to 1-jets coincide, i.e., $y_i^p = Y_i^p$. This construction extends to all orders and the semi-holonomic jets look in coordinates nearly as the holonomic ones, just losing the symmetry of the derivatives. See, e.g., [17] for detailed exposition.

$$\Omega^0(M, \mathbb{V}) \xrightarrow{\ d_{\mathbb{V}}\ } \Omega^1(M, \mathbb{V}) \xrightarrow{\ d_{\mathbb{V}}\ } \ \dots$$

$$\pi \Big\downarrow\Big\uparrow \mathrm{L} \qquad\qquad \pi \Big\downarrow\Big\uparrow \mathrm{L}$$

$$H_{\mathbb{V}}^0 M \qquad\qquad\quad H_{\mathbb{V}}^1 M$$

The quite straightforward idea is to seek for differential operators L, such that $d_{\mathbb{V}} \circ L$ are requested to be algebraically co-closed. Then the composition with the projection π makes sense and we could arrive at operators D

$$H_{\mathbb{V}}^k M \xdashrightarrow{\ D=\pi \circ d_{\mathbb{V}} \circ L\ } H_{\mathbb{V}}^{k+1} M.$$

The most important (and demanding) step in the original construction of the sequence of those operators in (2.49) was the following lemma in [10]. Notice [5] suggests a different and more efficient construction of these operators.

Lemma 2.3 *On each irreducible component of $H_{\mathbb{V}}^k M$, there is the unique strongly invariant operator L with values in* ker ∂^* *and splitting the projection π,*

$$H_{\mathbb{V}}^k M \underset{\pi}{\overset{L}{\rightleftarrows}} \Omega^k(M, \mathcal{V}),$$

such that $d_{\mathbb{V}} \circ L \in$ ker ∂^*.

The proof in [10] is very technical and there are many later improvements in the literature, starting with [5].

The resulting sequence of operators

$$H_{\mathbb{V}}^0 M \xdashrightarrow{\ D_0\ } H_{\mathbb{V}}^1 M \xdashrightarrow{\ D_1\ } H_{\mathbb{V}}^2 M \xdashrightarrow{\ D_2\ } \ \dots$$

is called the *BGG sequence* associated with the tractor bundle $\mathcal{V}$.

Theorem 2.8 *For each G-module $\mathbb{V}$, the BGG sequence is well defined on each Cartan geometry modelled on G/P and it restricts to the celebrated BGG resolution on the homogeneous model.*

If the twisted de-Rham sequence on a Cartan geometry is a complex, then also the BGG sequence is a complex, and they both compute the same cohomology of the underlying manifold.

A good example is the case when the Cartan geometry is torsion free and the curvature values act trivially on $\mathbb{V}$. Then the comparison (2.52) of the twisted exterior differential and the covariant exterior differential implies that the twisted de-Rham sequence will be exact.

Often only a part of the whole BGG sequence is exact and many celebrated complexes known in differential geometry can be recovered this way.

2.6.3   The First BGG Operators

Finally, we are coming to the first operators in BGG sequences. They are always overdetermined operators $D : H_V^0 M \to H_V^1 M$. Moreover, by the very construction, its space of solutions is in bijection with the space of the parallel tractors on the homogeneous model. Unfortunately, this is not true in general and the so-called *normal solutions* are those sections in the kernel of D which correspond to parallel tractors. See [9] for interesting results on the normal solutions. Because of lack of space in this last lecture, we shall just report briefly on the available results.

As carefully explained in [16], the normalization condition on the canonical tractor connections can be written as $\partial^*(R^V) = 0$, considered on the space of 2-forms valued in endomorphisms $V \otimes V^*$. At the same time, the normalization necessary for keeping the 1-1 correspondence between the solutions and the parallel tractors is rather $\partial_V^* R^V = 0$, where the codifferential is modified, see [16].

So, although the values of our operator L on the harmonic curvature are always algebraically co-closed, this is not enough.

The paper [16] answers positively the question: Can we modify the Cartan connection so that $\partial_V^* \circ d_V \circ L(\kappa) = 0$ and thus the 1-1 correspondence will hold true for all Cartan geometries?

The first useful observation is the fact that the BGG machinery construction survives without any changes if we restrict the deformations to the class of connections:

$$C = \{\tilde{\nabla} = \nabla + \Phi \mid \Phi \in \ker \partial_V^* \otimes \mathrm{id}_V, \ \Phi \text{ has homogeneity} \geqslant 1\}.$$

The main theorem of [16] says:

Theorem 2.9 *There is precisely one $\tilde{\nabla} \in C$ providing the 1-1 correspondence between* $\ker D_0$ *and* $\tilde{\nabla}$*-parallel tractors.*

At the very end, let us look again at the case of the "conformal to Einstein" Eq. (2.8), which is the first BGG operator for the choice of V equal to the conformal standard tractors T.

Clearly, $H_T^0 M$ is the projecting part $\mathcal{E}[1]$ of the tractors. Further, a straightforward check reveals that the operator (2.44) satisfies the conditions on L : $H_T^0 M \to \Omega^1(TM, TM)$. Indeed, the entire space of zero-forms is in the kernel of ∂^*, the exterior derivative d^ω is just the covariant derivative (2.19) of the tractor, its projecting slot vanishes, ∂^* maps the injecting slot to zero by the homogeneity, and ∂^* is given by the trace in the middle slot, which vanishes, too. Since the geometry is torsion free, the exterior covariant derivative coincides with the twisted derivative, see (2.52). Finally, the projection of $d_T \circ L$ to the harmonic component provides just the right operator (2.8) on $\mathcal{E}[1]$.

In this very special case, there is no need to modify the tractor connection in the above sense and thus there always is the 1-1 correspondence between the solutions and the parallel tractors, which is again realized directly by the operator L.

As already mentioned, many of important overdetermined operators appear as the first BGG operators. A vast supply of interesting examples of the first order ones appear in relation with the generalization of the classical problem of metrizability of a projective geometry into the realm of filtered manifolds and parabolic geometry. The projective case goes back to nineteenth century, the generalization was recently worked out in [7].

Acknowledgments The authors would like to thank the organizers of the Summer School in Wisła for the hospitality and great organization. The first author is also grateful to the audience for the patience and good questions.

The authors are most grateful to the anonymous referee who pointed out several gaps in the original manuscript.

The first author also acknowledges the support by the grant Nr. GA20-11473S of the Czech Grant Agency. The second author was supported by the grant MUNI/A/0885/2019 of the Masaryk University.

References

1. T.N. Bailey, M.G. Eastwood, A. Rod Gover, *Thomas's structure bundle for conformal, projective and related structures*, Rocky Mountain J. Math., 24 (1994), 1191–1217.
2. R.J. Baston, Almost Hermitian symmetric manifolds, I: Local twistor theory, II: Differential Invariants, Duke Math. J. 63 (1991), 81–111, 113–138.
3. R.J. Baston, M.G. Eastwood, The Penrose Transform: Its Interaction with Representation Theory, Courier Dover Publications, (2nd edition, 2016), 256pp.
4. I.N. Bernstein, I.M. Gelfand, S.I. Gelfand, Differential operators on the base affine space and a study of g-modules, in "Lie Groups and their Representations" (ed. I.M. Gelfand) Adam Hilger 1975, 21–64.
5. D.M.J. Calderbank, T. Diemer, *Differential invariants and curved Bernstein-Gelfand-Gelfand sequences*. J. Reine Angew. Math. 537 (2001), 67–103.
6. D.M.J. Calderbank, T. Diemer, V. Souček, *Ricci-corrected derivatives and invariant differential operators*, Diff. Geom. Appl. 23 (2005) 149–175.
7. D.M.J. Calderbank, J. Slovák, V. Souček, *Subriemannian metrics and the metrizability of parabolic geometries*, Journal of Geometric Analysis (2019), 32pp. (https://doi.org/10.1007/s12220-019-00320-1).
8. A. Čap, A.R. Gover, *Tractor calculi for parabolic geometries*, Trans. Amer. Math. Soc. 354 (2002), no. 4, 1511–1548.
9. A. Čap, A. R. Gover, Matthias Hammerl, *Normal BGG solutions and polynomials*, Internat. J. Math. 23 (2012), no. 11, 1250117, 29 pp.
10. A. Čap, J. Slovák, V. Souček, *Bernstein-Gelfand-Gelfand sequences*, Annals of Mathematics, Princeton University: The Johns Hopkins University Press, 2001, vol. 154, No 1, p. 97–113. (extended, more understandable version as ESI preprint No 722, see www.esi.ac.at).
11. A Čap, J. Slovák, *Parabolic Geometries I, Background and General Theory*, Providence, RI, USA: American Mathematical Society, 2009. 628 pp. Mathematical Surveys and Monographs, 154.
12. S. Curry, R. Gover, *An introduction to conformal geometry and tractor calculus, with a view to applications in general relativity*, arXiv:1412.7559v2, 74pp.
13. M.G. Eastwood, J. Slovák *Semiholonomic Verma modules*, Journal of Algebra, 1997, vol. 197, No 2, p. 424–448.

14. A.R. Gover, J. Slovák, *Invariant local twistor calculus for quaternionic structures and related geometries*, Journal of Geometry and Physics, Amsterdam: Elsevier Science, 1999, vol. 32, No 1, p. 14–56.

15. M. Hammerl, P. Somberg, V. Souček, J. Šilhan, *Invariant prolongation of overdetermined PDEs in projective, conformal, and Grassmannian geometry*, Ann. Global Anal. Geom. 42 (2012), no. 1, 121–145.

16. M. Hammerl, P. Somberg, V. Souček, J. Šilhan, *On a new normalization for tractor covariant derivatives*, J. Eur. Math. Soc. (JEMS) 14 (2012), no. 6, 1859–1883.

17. I. Kolář, P.W. Michor, J. Slovák, *Natural Operations in Differential Geometry*, Berlin-Heidelberg-New York: Springer-Verlag, 1993. 434 pp.

18. Bertrand Kostant, *Lie algebra cohomology and the generalized Borel-Weil theorem*, Ann. of Math. 74 (1961) 329–387.

19. J. Lepowsky, A generalization of the Bernstein-Gelfand-Gelfand resolution, J. of Algebra 49 (1977), 496–511.

20. J. Slovák, *Natural operators on conformal manifolds*, Research Lecture Notes, University of Vienna, 1992, 138pp., (www.math.muni.cz/~slovak/Papers/habil.pdf)

21. J. Slovák, *Parabolic Geometries*, Research Lecture Notes, Adelaide, 1997, 70pp., (www.math.muni.cz/~slovak/Papers/examples.pdf)

22. T.Y. Thomas, *On conformal geometry*. Proc. N.A.S. 12 (1926), 352–359.

23. V. Wünsch, *On Conformally Invariant Differential Operators*, Mathematische Nachrichten, 129 (1986), 269–281.

24. R. Zierau, Representations in Dolbeault Cohomology, in Representation Theory of Lie Groups edited by Jeffrey Adams, David Vogan, IAS/Park City Mathematics Series, American Math. Soc., vol. 8 (2015), 89–146.

Chapter 3
Symmetries and Integrals

Valentin V. Lychagin

3.1 Preface

In these lectures, I want to illustrate an application of symmetry ideas to integration
of differential equations. Basically, we will consider only differential equations of
finite type, i.e. equations with finite-dimensional space Sol of (local) solutions. Ordinary differential equations make up one of the main examples of such equations. The
symmetry Lie algebra Sym induces an action on manifold Sol. In the case when this
action is transitive, we expect to get more detailed information on solutions. Here,
we are going to realize this expectation; namely, we will show that in the case when
the Lie algebra Sym is solvable, integration of the differential equation can be done
by quadratures due to the Lie–Bianchi theorem (see, for example, [4] or [6]). In
the case when the Lie algebra Sym contains simple subalgebras, integration shall
use quadratures (for radical of the Lie algebra) and integration of some differential
equations, which we will call *model* equations [6, 10]. The model equations depend
on the type of the simple Lie subalgebras and are natural generalizations of the well-
known Riccati differential equations. They possess nonlinear *Lie superposition*,
and all their solutions could be obtained by nonlinear superposition of a finite set
of solutions (the so-called *fundamental solutions*). Once more, the form of this
superposition and the number of fundamental solutions are dictated by the symmetry
Lie algebra. In order to give a more "practical" reader a feeling of the power of
the geometrical approach to differential equations, we included in these lectures a
number of examples on the formula level.

The paper is organized as follows. First, we consider symmetries of two types
of distributions: Cartan distributions and completely integrable distributions. We

V. V. Lychagin (✉)
Institute of Control Sciences of Russian Academy of Sciences, Moscow, Russia
e-mail: valentin.lychagin@uit.no

© The Author(s), under exclusive license to Springer Nature Switzerland AG 2021   73
M. Ulan, E. Schneider (eds.), *Differential Geometry, Differential Equations, and
Mathematical Physics*, Tutorials, Schools, and Workshops in the Mathematical
Sciences, https://doi.org/10.1007/978-3-030-63253-3_3

show how to integrate completely integrable distributions possessing a symmetry
Lie algebra that acts in a transitive way on the space of (local) integral manifolds.
We begin with the most trivial (but also the more applicable) case of a commutative
symmetry Lie algebra and show how to get quadratures. As an example of
application of this case, we discuss the famous Liouville–Arnold theorem on
integrable Hamilton systems. We propose also the general reduction principle that
allows to split integration for pair $(\mathcal{J}, \text{Sym})$, where $\mathcal{J} \subset \text{Sym}$ is an ideal in
the symmetry Lie algebra, into two separate cases with symmetry algebra $\mathcal{J}$ and
$\text{Sym}/\mathcal{J}$. Application of this principle to solvable Lie algebras gives us the Lie–
Bianchi theorem (together with a constructive method of finding quadratures), as
well as integration of cases with general symmetry algebra Sym by means of
model equations. We conclude these lectures by showing applications to ordinary
differential equations and especially to the "toy" case of Schrödinger type equations:
$y'' + W(x)y = 0$. We show that (nontrivial) geometries on the line are hidden in
these equations and how symmetries allow us to write explicit solutions of these
equations.

I consider these lectures as an invitation to the wonderful world of symmetries
and differential equations. More details, results, and methods can be found in more
advanced expositions (see, for example, the cited books at the end of the lectures).

3.2   Distributions

Let M be an $(n + m)$-dimensional smooth manifold, and let $\tau : TM \to M$ be the
tangent bundle. Then, a distribution P on M is a smooth field

$$P : a \in M \longmapsto P(a) \subset T_a M$$

of m-dimensional subspaces of the tangent spaces. The number m is called the
dimension of the distribution, $m = \dim P$, and the number n is called a *codimension*
of P, $n = \text{codim} P$.

The statement that P is a smooth family could be formulated in two different
ways:

1. For any point $a \in M$, there are vector fields $\langle X_1, \ldots, X_m \rangle$ defined in a
   neighborhood O of a such that the vectors $X_{i,b} \in T_b M$, $i = 1, \ldots, m$ belong
   to and form a basis in $P(b)$ for every $b \in O$.
2. There are differential 1-forms $\langle \omega_1, \ldots, \omega_n \rangle$, such that

$$P(b) = \ker \omega_{1,b} \cap \cdots \cap \ker \omega_{n,b}$$

in the neighborhood O.

For the first definition, let us introduce $C^\infty (M)$-module

$$D(P) = \{X \in \text{Vect}(M), X_a \in P(a), \forall a \in M\}$$

of all smooth vector fields lying in (or tangent to) the distribution P.

Then, (1) states that this module is locally free (=projective) and that the sets $\langle X_1, \ldots, X_m \rangle$ give us local bases for $D(P)$. In a similar way, let us introduce another $C^\infty (M)$-module

$$\text{Ann}(P) = \left\{\omega \in \Omega^1 (M), \omega(X) = 0, \forall X \in D(P)\right\}$$

of all smooth differential 1-forms vanishing on vector fields from P. Then, (2) states that this module is also locally free and that the sets $\langle \omega_1, .., \omega_n \rangle$ are local bases for this module.

Example 3.1 Consider $M = \mathbb{R}^3$ with coordinates (x, y, z), and let $\omega = dz - y dx$. This form does not vanish at any point of M and therefore defines a distribution P on M of dimension 2 and codimension 1. The vector fields

$$X_1 = \frac{\partial}{\partial y}, \quad X_2 = \frac{\partial}{\partial x} + y \frac{\partial}{\partial z} \tag{3.1}$$

give us a basis in the module $D(P)$.

A submanifold $N \subset M$ is said to be *integral* for the distribution P if

$$T_a N \subset P(a), \tag{3.2}$$

for all $a \in N$. This condition is better to formulate in terms of the differential 1-forms $\langle \omega_1, .., \omega_n \rangle$. Then, N is integrable if their restrictions to N are equal to zero:

$$\omega_i |_N = 0, \quad i = 1, \ldots, n.$$

An integral manifold is *maximal* if it is not contained in an integral manifold of greater dimension.

Example 3.2 The distribution (3.1) has 1-dimensional integral submanifolds. Namely, assume that x is a coordinate on N, i.e.

$$N = \{z = A(x), \ y = B(x)\},$$

for some smooth functions A and B. Then,

$$\omega|_N = dA - B dx = (A' - B) dx,$$

and N is integral if and only if $B = A'$, and

$$N = \left\{ z = A(x), \; y = A'(x) \right\}. \tag{3.3}$$

On the other hand, if N is a 2-dimensional integral manifold, then the vector field $\partial/\partial z$ is not tangent to N, because $\partial/\partial z \notin D(P)$. Therefore, we can represent

$$N = \{z = A(x, y)\}.$$

Then, $\omega|_N = dA - ydx = (A_x - y)\,dx + A_ydy = 0$ if and only if

$$A_x = y, \; A_y = 0, \tag{3.4}$$

which is impossible.

Two observations should be made from this example: (1) maximal integral manifolds can have dimension less than dim P and (2) finding of integral manifolds is equivalent to finding of solutions of some differential equations.

A distribution is said to be *completely integrable* if the dimension of every maximal integral manifold is exactly the dimension of the distribution itself, and if for any point of M, there is a maximal integral manifold containing this point. For such distributions, the entire manifold can be presented as the disjoint union of maximal integral manifolds of the distribution, which are the leaves of a foliation, so that the notion of a completely integrable distribution is equivalent to that of a foliation.

Theorem 3.1 (Frobenius) *A distribution P is completely integrable if and only if the module $D(P)$ is closed with respect to commutator of vector fields*

$$X, Y \in D(P) \Longrightarrow [X, Y] \in D(P). \tag{3.5}$$

Moreover, if the distribution P is completely integrable, and if N_1 and N_2 are integral submanifolds of P, passing through a point $a \in N_1 \cap N_2$, then $N_1 = N_2$ in a neighborhood of the point.

Example 3.3 Consider the distribution given by (3.1). As we have seen, the maximal integral manifolds for this distribution have dimension 1, and therefore this distribution is not completely integrable. On the other hand, the module $D(P)$ for this distribution is generated by the vector fields

$$X_1 = \frac{\partial}{\partial y}, \quad X_2 = \frac{\partial}{\partial x} + y\frac{\partial}{\partial z},$$

and we have

$$[X_1, X_2] = \frac{\partial}{\partial z} \notin D(P).$$

The condition (3.5) can be reformulated in terms of differential forms generating the module Ann (P). Namely, using the formula

$$d\omega\,(X, Y) = X\,(\omega\,(Y)) - Y\,(\omega\,(X)) - \omega\,([X, Y]),$$

we get

$$d\omega\,(X, Y) = -\omega\,([X, Y]),$$

for all $X, Y \in D\,(P)$ and $\omega \in \mathrm{Ann}\,(P)$. Therefore, the condition (3.5) is equivalent to

$$d\omega\,(X, Y) = 0,$$

for all $X, Y \in D\,(P)$. In other words, the restriction of the differential 2-form $d\omega|_P$ on the distribution P vanishes for all forms $\omega \in \mathrm{Ann}\,(P)$.

In terms of local bases $\langle X_1, \ldots, X_m \rangle$ for the module $D\,(P)$ or $\langle \omega_1, .., \omega_n \rangle$ for the module $\mathrm{Ann}\,(P)$, the conditions for complete integrability can be reformulated in the following equivalent forms:

1.

$$[X_i, X_j] = \sum_k c_{ij}^k X_k, \tag{3.6}$$

for all $i, j = 1, \ldots, m$ and some smooth functions c_{ij}^k.

2.

$$d\omega_i = \sum_j \gamma_{ij} \wedge \omega_j, \tag{3.7}$$

for all $i, j = 1, \ldots, n$ and some differential 1-forms γ_{ij}.

3.3   Distributions and Differential Equations

3.3.1   Cartan Distributions (ODEs)

Let $\mathbf{J}^k$ be the space of all k-jets of functions in one variable x. Then, the k-jet of a smooth function $f(x)$ at point $a \in \mathbb{R}$, denoted by $[f]_a^k$, is given by its set of derivatives:

$$[f]_a^k = \left(f\,(a), f'\,(a), \ldots, f^{(k)}\,(a) \right).$$

We denote by $(x, u_0, u_1, \ldots, u_k)$ the coordinates on $\mathbf{J}^k$ satisfying

$$u_i\left([f]_a^k\right) = f^{(i)}(a),$$

for all $i = 0, 1, \ldots k$, and $x\left([f]_a^k\right) = a$.

The differential 1-forms

$$\omega_i = du_i - u_{i+1}dx, \tag{3.8}$$

on $\mathbf{J}^k$, where $i = 0, \ldots, k-1$, we call *Cartan forms*, and the distribution

$$C_k = \ker \omega_0 \cap \cdots \cap \ker \omega_{k-1}$$

we call the *Cartan distribution* on the jet-space.

We have $\dim \mathbf{J}^k = k + 2$, $\operatorname{codim} C_k = k$, and therefore $\dim C_k = 2$, i.e. C_k is a distribution of planes.

It is easy to see that a basis in the module $D(C_k)$ of vector fields tangent to C_k is formed by vector fields

$$X_1 = \frac{\partial}{\partial u_k}, \tag{3.9}$$

$$X_2 = \frac{\partial}{\partial x} + u_1 \frac{\partial}{\partial u_0} + \cdots + u_k \frac{\partial}{\partial u_{k-1}},$$

and

$$[X_1, X_2] = \frac{\partial}{\partial u_{k-1}} \notin D(C_k).$$

Therefore, C_k is a non-integrable distribution, and its maximal integral manifolds have dimension 1.

To find these curves, we remark that trajectories of the vector field X_1 are integral curves for C_k and, similarly to (3.1), integral curves N on which x is a coordinate have the special form

$$N = L_A^{(k)} = \left\{ u_0 = A(x), u_1 = A'(x), \ldots, u_k = A^{(k)}(x) \right\}.$$

By an *ordinary differential equation* (ODE) of order k, we mean a relation that connects components of k-jets $[f]_a^k$ of unknown functions $f(x)$, i.e. a relation of the form

$$F(x, u_0, \ldots, u_k) = 0, \tag{3.10}$$

which is valid when u_i are coordinates of solutions.

Geometrically, it means that we consider a submanifold (possibly with singularities)

$$\mathcal{E}_F = \{F(x, u_0, \ldots, u_k) = 0\} \subset \mathbf{J}^k,$$

and solutions are curves

$$L_A^{(k)} \subset \mathcal{E}_F.$$

Assuming that $\mathcal{E}_F$ is a smooth submanifold of $\mathbf{J}^k$, we say that solutions of the ODE $\mathcal{E}_F$ are integral curves of the Cartan distribution C_k lying in $\mathcal{E}_F$, or, in other words, they are integral curves of the restriction of the distribution C_k on $\mathcal{E}_F$:

$$C_F : a \in \mathcal{E}_F \rightarrow C_k(a) \cap T_a(\mathcal{E}_F).$$

Remark that $\dim C_F(a) = 1$ if $C_k(a)$ is not a subspace of $T_a(\mathcal{E}_F)$ and that $\dim C_F(a) = 2$ if $C_k(a) \subset T_a(\mathcal{E}_F)$. In the last case, we say that point $a \in \mathcal{E}_F$ is a *singular point*.

Therefore, on the complement $\mathcal{E}_F \setminus \text{Sing}(\mathcal{E}_F)$, we have a 1-dimensional distribution C_F. This is obviously a completely integrable distribution, and its integral curves $L \subset \mathcal{E}_F \setminus \text{Sing}(\mathcal{E}_F)$ are smooth solutions of the equation if and only if function x is a coordinate on L.

To find a basis X in the module $D(C_F)$, we write down vector field X in the form

$$X = a(x)X_1 + b(x)X_2,$$

where X_1 and X_2 form basis in C_k. Then, $X \in D(C_F)$ if and only if X is tangent to $\mathcal{E}_F$ or if

$$X(F) = 0,$$

on $\mathcal{E}_F$. Thus,

$$a\frac{\partial F}{\partial u_k} + bD_k(F) = 0$$

on $\mathcal{E}_F$. Here, we denoted by

$$D_k = \frac{\partial}{\partial x} + u_1\frac{\partial}{\partial u_0} + \cdots + u_k\frac{\partial}{\partial u_{k-1}} \qquad (3.11)$$

the vector field X_2 in (3.9).

Remark that $\text{Sing}(\mathcal{E}_F)$ is given by equations

$$\frac{\partial F}{\partial u_k} = 0, \ D_k\,(F) = 0, \ F = 0$$

and in general defines a submanifold $\mathrm{Sing}\,(\mathcal{E}_F) \subset \mathcal{E}_F$ of codimension 2.

Equations $\mathcal{E}_F$, where $\frac{\partial F}{\partial u_k} \neq 0$, are called equations of *principal type*, and for such equations, a basis in $D\,(C_F)$ has the form

$$D = \frac{\partial F}{\partial u_k} D_k - D_k\,(F)\,\frac{\partial}{\partial u_k},$$

and x is a coordinate on integral curves.

Thus, for principal type equations, solutions are smooth functions, and in all other cases, they are singular and multivalued functions (see [7] for more such examples).

To simplify the formulae in what follows, we will consider only principal type equations of the form

$$u_k = F\,(x, u_0, \ldots, u_{k-1})\,. \tag{3.12}$$

Then, $\mathcal{E}_F$ is diffeomorphic to $\mathbf{J}^{k-1}$, $(x, u_0, \ldots, u_{k-1})$ are coordinates on $\mathcal{E}_F$, and the basic vector field in $D\,(C_F)$ has the form

$$D_k = \frac{\partial}{\partial x} + u_1 \frac{\partial}{\partial u_0} + \cdots + F \frac{\partial}{\partial u_{k-1}}.$$

The distribution C_F can also be defined by the following Cartan forms:

$$\omega_0 = du_0 - u_1 dx, \tag{3.13}$$

$$\ldots\ldots\ldots\ldots$$

$$\omega_{k-2} = du_{k-2} - u_{k-1} dx,$$

$$\omega_{k-1} = du_{k-1} - F dx.$$

When working with algebraic equations, we use different algebraic manipulations in order to simplify them. For differential equations, the class of possible manipulations can be essentially extended by adding the operation of differentiation or *prolongation*.

Take, for example, system (3.4)

$$\frac{\partial A}{\partial x} = y,$$

$$\frac{\partial A}{\partial y} = 0,$$

which we investigated above. Then, by differentiating, the first and second equations in x and y, we get the following system:

$$\frac{\partial A}{\partial x} = y, \ \frac{\partial^2 A}{\partial x^2} = 0, \ \frac{\partial^2 A}{\partial x \partial y} = 1,$$

$$\frac{\partial A}{\partial y} = 0, \ \frac{\partial^2 A}{\partial x \partial y} = 0, \ \frac{\partial^2 A}{\partial y^2} = 0,$$

which is obviously contradictory and therefore has no solutions.

To apply the prolongation procedure to the ordinary differential equations, we introduce the formal derivation (the *total derivative* in x)

$$D = \frac{\partial}{\partial x} + u_1 \frac{\partial}{\partial u_0} + \cdots + u_k \frac{\partial}{\partial u_{k-1}} + u_{k+1} \frac{\partial}{\partial u_k} + \cdots.$$

Then, the prolongation of Eq. (3.10) is the following system $\mathcal{E}_F^{(1)} \subset \mathbf{J}^{k+1}$:

$$F(x, u_0, \ldots, u_k) = 0,$$

$$D(F) = D_k(F) + u_{k+1} \frac{\partial F}{\partial u_k} = 0.$$

Applying in series this procedure, we get l-th prolongations $\mathcal{E}_F^{(l)} \subset \mathbf{J}^{k+l}$ given by relations

$$F = 0, \ D(F) = 0, \ D^2(F) = 0, \ldots, \ D^l(F) = 0. \tag{3.14}$$

The advantage of using prolongations $\mathcal{E}_F^{(l)}$ and their inverse limit $\mathcal{E}_F^{(\infty)} \subset \mathbf{J}^{\infty}$ comes from the fact that these equations contain information on all derivatives of solutions up to order l or ∞.

It is easy to see that the Cartan distributions C on the prolongations are still 1-dimensional at regular points and generated by the restrictions of the total derivative D on $\mathcal{E}_F^{(l)}$.

3.3.2 Cartan Distributions (PDEs)

For the case of functions of n variables $x = (x_1, \ldots, x_n)$, and corresponding partial differential equations, the above constructions can be repeated, practically word by word.

Namely, denote by $\mathbf{J}^k(n)$ the space of all k-jets of functions in n variables. Then, the k-jet $[f]_a^k$ of a smooth function $f(x_1, \ldots, x_n)$ at a point $a \in \mathbb{R}^n$ is given by the values of its derivatives

$$\frac{\partial^\sigma f}{\partial x^\sigma}(a)$$

at the point. Here, $\sigma = (\sigma_1, \ldots, \sigma_n)$ are multi-indices of order $0 \leqslant |\sigma| \leqslant k$, where $|\sigma| = \sigma_1 + \cdots + \sigma_n$.

Denote by $(x, u_\sigma, 0 \leqslant |\sigma| \leqslant k)$ the standard coordinates on $\mathbf{J}^k(n)$, where

$$u_\sigma\left([f]_a^k\right) = \frac{\partial^\sigma f}{\partial x^\sigma}(a).$$

Define also the Cartan forms

$$\omega_\sigma = du_\sigma - \sum_i u_{\sigma+1_i} dx_i$$

and the Cartan distribution

$$C_k = \bigcap_{0 \leqslant |\sigma| \leqslant k-1} \ker \omega_\sigma.$$

We have $\dim \mathbf{J}^k = n + \binom{n+k}{k}$, $\operatorname{codim} C_k = \binom{n+k-1}{k-1}$, and therefore $\dim C_k = n + \binom{n+k-1}{k}$.

The following vector fields make up a basis in the module $D(C_k)$:

$$X_\sigma = \frac{\partial}{\partial u_\sigma}, |\sigma| = k,$$

$$Y_i = \frac{\partial}{\partial x_i} + \sum_{|\sigma| \leqslant k-1} u_{\sigma+1_i} \frac{\partial}{\partial u_\sigma}.$$

We have $[X_\sigma, Y_i] = \frac{\partial}{\partial u_{\sigma-1_i}} \notin D(C_k)$ if $\sigma_i \geqslant 1$, and therefore the Cartan distribution is not completely integrable.

Similarly to the 1-dimensional case, this distribution has two types maximal integral manifolds:

1.

$$L_A^{(k)} = \left\{ u_0 = A(x), \ u_\sigma = \frac{\partial^{|\sigma|} A}{\partial x^\sigma} \right\},$$

where $A(x)$ is a smooth function, and
2. integral manifolds of the completely integrable distribution generated by all vector fields X_σ.

Remark that the dimension of the first type of integral manifolds equals n, while the dimension for the second type of integral manifolds equals $\binom{n+k-1}{k}$, and $\binom{n+k-1}{k} > n$ if $n \geqslant 1$ and $k \geqslant 2$. For the complete description of various maximal integral manifolds (and their dimensions) for the Cartan distributions, see [11].

As above, we will consider differential equations (PDEs in this case) as submanifolds

$$\mathcal{E}_F = \{F(x, u_\sigma) = 0\} \subset \mathbf{J}^k(n)$$

and their smooth solutions as submanifolds $L_A^{(k)} \subset \mathcal{E}_F$.

In the more general case, we will by solution mean any such n-dimensional submanifold L of the Cartan distribution that $L \subset \mathcal{E}_F$, i.e. n-dimensional integral submanifold L of the restriction of the Cartan distribution on $\mathcal{E}_F$.

By using the total derivations

$$D_i = \frac{\partial}{\partial x_i} + \sum u_{\sigma+1_i} \frac{\partial}{\partial u_\sigma},$$

where $i = 1, .., n$, we define prolongations

$$\mathcal{E}_F^{(l)} = \left\{F(x, u_\sigma) = 0, \ D^\sigma(F) = 0, \ |\sigma| \leqslant l\right\} \subset \mathbf{J}^{k+l}(n),$$

which contains all information about $(k + l)$-jets of solutions.

3.4 Symmetry

By a (*finite*) *symmetry* of the distribution P on the manifold M, we understand a (possibly local) diffeomorphism $\phi : M \to M$, which takes P into itself, i.e. such that $\phi_*(P_a) = P_{\phi(a)}$, for all points $a \in M$, or, in short, $\phi_*(P) = P$.

A vector field X is said to be (*an infinitesimal*) *symmetry* of the distribution if the flow generated by X consists of finite symmetries.

The infinitesimal approach turns out to be much more constructive and more algebraic than its finite counterpart, so in what follows the word symmetry will always mean infinitesimal symmetry unless otherwise explicitly specified.

Assume that distribution P is generated by differential 1-forms $\omega_1, \ldots, \omega_m$, where $m = \text{codim } P$. We write $P = \langle \omega_1, \ldots, \omega_m \rangle$. Then, the condition $\phi_*(P) = P$ means that the differential 1-forms $\phi^*(\omega_1), \ldots, \phi^*(\omega_m)$ determine the same distribution P and therefore can be expressed in terms of the basis forms

$$\phi^*(\omega_1) = a_{11}\omega_1 + \cdots + a_{1m}\omega_m,$$

$$\ldots\ldots\ldots\ldots$$

$$\phi^*(\omega_m) = a_{m1}\omega_1 + \cdots + a_{mm}\omega_m,$$

for some smooth functions a_{ij}, or in the equivalent form

$$\phi^*(\omega_1) \wedge \omega_1 \wedge \cdots \wedge \omega_m = 0, \ \ldots, \ \phi^*(\omega_m) \wedge \omega_1 \wedge \cdots \wedge \omega_m = 0. \tag{3.15}$$

These conditions take the form

$$L_X(\omega_1) = a_{11}\omega_1 + \cdots + a_{1m}\omega_m,$$

$$\ldots\ldots\ldots\ldots\ldots$$

$$L_X(\omega_m) = a_{m1}\omega_1 + \cdots + a_{mm}\omega_m,$$

or

$$L_X(\omega_1) \wedge \omega_1 \cdots \wedge \omega_m = 0, \quad \ldots, \quad L_X(\omega_m) \wedge \omega_1 \cdots \wedge \omega_m = 0, \qquad (3.16)$$

for infinitesimal symmetries X.

Let us denote by $\mathrm{Sym}(P)$ the set of all infinitesimal symmetries of the distribution P. Then, the above formulae show that the following conditions are equivalent:

- $X \in \mathrm{Sym}(P)$,
- $Y \in D(P) \Longrightarrow [X, Y] \in D(P)$,
- $\omega \in \mathrm{Ann}(P) \Longrightarrow L_X(\omega) \in \mathrm{Ann}(P)$.

It follows that $\mathrm{Sym}(P)$ is a Lie algebra over $\mathbb{R}$ with respect to the commutator of vector fields.

3.4.1 Symmetries of the Cartan Distributions

Let us now consider symmetries of the Cartan distributions first on $\mathbf{J}^k$.

Lemma 3.1 *We have*

$$df = D_k(f)\, dx + \frac{\partial f}{\partial u_k} du_k \mod \langle \omega_0, \ldots, \omega_{k-1} \rangle,$$

for any smooth function f on $\mathbf{J}^k$.

Proof Indeed, we have

$$du_i = u_{i+1} dx + \omega_i,$$

for all $i = 0, \ldots, k - 1$. Therefore,

$$df = \frac{\partial f}{\partial x} dx + \sum_{i=0}^{k} \frac{\partial f}{\partial u_i} du_i = \frac{\partial f}{\partial x} dx + \sum_{i=0}^{k-1} u_{i+1} \frac{\partial f}{\partial u_i} dx + \frac{\partial f}{\partial u_k} du_k + \sum_{i=0}^{k-1} \frac{\partial f}{\partial u_i} \omega_i$$

$$= D_k(f)\, dx + \frac{\partial f}{\partial u_k} du_k \mod \langle \omega_0, \ldots, \omega_{k-1} \rangle.$$

$\square$

Now, let $X \in \text{Sym}(C_k)$ and

$$X = a\frac{\partial}{\partial x} + \sum_{i=0}^{k} A_i \frac{\partial}{\partial u_i},$$

where a and A_i are smooth functions on $\mathbf{J}^k$. Then, by using the above lemma, we get

$$L_X(\omega_i) = dA_i - u_{i+1}da - A_{i+1}dx =$$

$$(D_k(A_i) - u_{i+1}D_k(a) - A_{i+1})dx + \left(\frac{\partial A_i}{\partial u_k} - u_{i+1}\frac{\partial a}{\partial u_k}\right)du_k \mod \langle\omega_0, \ldots, \omega_{k-1}\rangle,$$

for $i = 0, \ldots, k - 1$. Therefore,

$$A_{i+1} = D_k(A_i) - u_{i+1}D_k(a), \tag{3.17}$$

$$\frac{\partial A_i}{\partial u_k} - u_{i+1}\frac{\partial a}{\partial u_k} = 0,$$

for $i \leqslant k - 1$.

Let us introduce the functions

$$\phi_i = A_i - u_{i+1}a,$$

for all $i \leqslant k - 1$. Then, the system (3.17) takes the form:

$$\phi_{i+1} = D_k(\phi_i),$$

$$\frac{\partial \phi_i}{\partial u_k} = 0,$$

for $i \leqslant k - 2$, and

$$A_k = D_k(\phi_{k-1}),$$

$$a = -\frac{\partial \phi_{k-1}}{\partial u_k}.$$

Therefore,

$$\phi_i = D_k^i(\phi_0),$$

for $i \leqslant k - 2$, and the condition $\frac{\partial \phi_{k-2}}{\partial u_k} = 0$ implies that $\phi_0 = \phi(x, u_0, u_1)$, and

$$a = -\frac{\partial \phi_{k-1}}{\partial u_k} = -\frac{\partial \phi}{\partial u_1}.$$

Moreover,

$$A_i = D_k^i(\phi) + \frac{\partial \phi}{\partial u_1} u_{i+1},$$

for $i \leqslant k - 1$, and

$$A_k = D_k^k(\phi).$$

Summarizing, we get the following description of symmetries of the Cartan distribution.

Theorem 3.2 (Bäcklund–Lie) *Any symmetry $X \in \mathrm{Sym}(C_k)$ has the form*

$$X = \sum_{i=0}^{k} D_k^i(\phi) \frac{\partial}{\partial u_i} - \frac{\partial \phi}{\partial u_1} D_k, \tag{3.18}$$

for some smooth function $\phi = \phi(x, u_0, u_1)$.

Remark 3.1

1. We call the function ϕ, which defines the symmetry, the *generating function*, and the corresponding vector field X will be denoted by X_ϕ. Thus,

$$\phi = \omega_0(X_\phi).$$

2. The commutator of two symmetries X_ϕ and X_ψ is also a symmetry. Denote its generating function by $[\phi, \psi]$, then

$$[\phi, \psi] = \omega_0([X_\phi, X_\psi]),$$

and the bracket $[\phi, \psi]$ (called the *Lagrange bracket*) defines a Lie algebra structure on $C^\infty(\mathbf{J}^1)$.

Moreover, a straightforward computation shows that

$$[\phi, \psi] = X_\phi(\psi) - X_1(\phi)\psi. \tag{3.19}$$

3. The Cartan distribution C_1 defines the contact structure on $\mathbf{J}^1$. The elements of $\mathrm{Sym}(C_1)$ are called contact vector fields, and they also have form (3.18)

$$X_\phi = \phi \frac{\partial}{\partial u_0} + D_1(\phi) \frac{\partial}{\partial u_1} - \frac{\partial \phi}{\partial u_1} D_1, \tag{3.20}$$

where

$$D_1 = \frac{\partial}{\partial x} + u_1 \frac{\partial}{\partial u_0}.$$

4. Vector fields (3.18) are prolongations of (3.20).

Similar results are valid for symmetries of Cartan distributions in the jet spaces $\mathbf{J}^k(n)$.

Theorem 3.3 (Lie–Bäcklund) *Any symmetry* $X \in \mathrm{Sym}(C_k)$ *on* $\mathbf{J}^k(n)$ *has the form*

$$X_\phi = \sum_{|\sigma| \leqslant k} D_k^\sigma(\phi) \frac{\partial}{\partial u_\sigma} - \sum_{i=1}^n \frac{\partial \phi}{\partial u_i} D_{i,k}, \qquad (3.21)$$

where $\phi(x_1, \ldots, x_n, u_0, u_1, \ldots, u_n) = \omega_0(X_\phi)$ *is the generating function, and*

$$D_{i,k} = \frac{\partial}{\partial x_i} + u_i \frac{\partial}{\partial u_0} + \cdots + u_{\sigma+1_i} \frac{\partial}{\partial u_\sigma} + \cdots, \quad |\sigma| \leqslant k - 1,$$
$$D_k^\sigma = D_{1,k}^{\sigma_1} \circ \cdots \circ D_{n,k}^{\sigma_n}.$$

The Lagrange bracket $[\phi, \psi] = X_\phi(\psi) - X_1(\phi)\psi$ *defines the Lie algebra structure on* $C^\infty(\mathbf{J}^1(n))$, *and, as above,* $[\phi, \psi] = \omega_0([X_\phi, X_\psi])$.

3.4.2 Symmetries of Completely Integrable Distributions

Let $X \in \mathrm{Sym}(P)$ be a symmetry of a completely integrable distribution P, and let $A_t : M \to M$ be the corresponding flow. Then, for any integral manifold $L \subset M$, the submanifolds $A_t(L)$ are also integral.

In other words, a symmetry X generates a flow on the set $\mathrm{Sol}(P)$ of all maximal integral manifolds. There is, however, a distinguished class of symmetries (they are called *characteristic symmetries*) which generate trivial flows, i.e. they leave invariant every integral manifold. Namely, we have $[D(P), D(P)] \subset D(P)$, because P is a completely integrable distribution, and therefore $D(P) \subset \mathrm{Sym}(P)$.

Moreover, the vector fields $X \in D(P)$ are tangent to any maximal integral manifold of P and therefore generate the trivial flow on $\mathrm{Sol}(P)$. The relation $[\mathrm{Sym}(P), D(P)] \subset D(P)$ shows that $D(P)$ is an ideal in the Lie algebra $\mathrm{Sym}(P)$. We call elements of the quotient Lie algebra

$$\mathrm{Shuf}(P) = \mathrm{Sym}(P)/D(P)$$

shuffling symmetries of the distribution P.

The name reflects the fact that flows on Sol (P) corresponding to different representatives of a class $X \mod D(P)$ rearrange, or shuffle, the set of maximal integral manifolds of P in the some way.

3.5   The Lie–Bianchi Theorem

Let P be a completely integrable distribution, codim $P = m$, generated by the differential 1-forms $\omega_1, \ldots, \omega_m$, and let $\mathfrak{g} \subset \operatorname{Shuf}(P)$ be a Lie subalgebra with $\dim \mathfrak{g} = m$. Let $\overline{X}_1, \ldots, \overline{X}_m$ be a basis of $\mathfrak{g}$, where $\overline{X}_i = X_i \mod D(P)$ for $X_i \in \operatorname{Sym}(P), i = 1, \ldots, m)$.

Suppose that $\mathfrak{g}$ is *transversal* to the distribution in the sense that the natural mappings

$$\mathfrak{g} \to T_a(M) / P_a$$

are isomorphisms for all points $a \in M$.

The problem of integration of a distribution consists of describing its maximal integral manifolds. For completely integrable distributions, this is equivalent to finding a complete set of first integrals.

A function $H \in C^\infty(M)$ is called a *first integral of the distribution* P if every integral manifold of P lies entirely in some level surface $M_c = H^{-1}(c)$ or, equivalently, if $Z(H) = 0$ for every vector field $Z \in D(P)$ or, equivalently, if $dH \in \operatorname{Ann}(P)$.

A *complete set of first integrals* of the distribution P is a set of functions $H_1, \ldots, H_m$ with the property that

$$M_{c_1, \ldots, c_m} = H_1^{-1}(c_1) \cap \cdots \cap H_m^{-1}(c_m)$$

represent the set of all maximal integral manifolds of P in some dense domain of M. In this section, we discuss a method to find such integrals when we have a transversal shuffling algebra of symmetries.

3.5.1   *Commutative Lie Algebra Symmetries*

Assume that the transversal shuffling algebra of symmetries $\mathfrak{g}$ is commutative. The matrix

$$\Xi = \begin{Vmatrix} \omega_1(X_1) & \cdots & \omega_1(X_m) \\ \vdots & \vdots & \vdots \\ \omega_m(X_1) & \cdots & \omega_m(X_m) \end{Vmatrix}$$

is nondegenerate: $\det(\Xi) \neq 0$.

Let us choose another basis $\omega'_1, .., \omega'_m$ in the module Ann(P) such that

$$\omega'_i(X_j) = \delta_{ij}. \tag{3.22}$$

Indeed, we have relations

$$\omega'_1 = a_{11}\omega_1 + \cdots + a_{1m}\omega_m,$$

$$\cdots\cdots\cdots\cdots\cdots$$

$$\omega'_m = a_{m1}\omega_1 + \cdots + a_{mm}\omega_m,$$

where the matrix $A = \|a_{ij}\|$ is also nondegenerate.

The condition (3.22) is equivalent to $A\varXi = 1$, or $A = \varXi^{-1}$. In other words, the differential forms $\omega'_1, .., \omega'_m$, where

$$\begin{Vmatrix} \omega'_1 \\ \vdots \\ \omega'_m \end{Vmatrix} = \varXi^{-1} \begin{Vmatrix} \omega_1 \\ \vdots \\ \omega_m \end{Vmatrix},$$

satisfy condition (3.22). To simplify notation, let us assume that the basis $\omega_1, .., \omega_m$ is normalized from the very beginning, i.e. satisfies condition (3.22).

Then, we have

$$d\omega_i(X_k, X_l) = X_k(\omega_i(X_l)) - X_l(\omega_i(X_k)) - \omega_i([X_k, X_l]) = 0,$$

because of the commutativity of Lie algebra $\mathfrak{g}$ and condition (3.22).

We also have

$$d\omega_i(X_k, Z) = X_k(\omega_i(Z)) - Z(\omega_i(X_k)) - \omega_i([X_k, Z]) = 0,$$

for all vector fields $Z \in D(P)$, because $\omega_i(Z) = 0$ and $[X_k, Z] \in D(P)$.

And finally

$$d\omega_i(Z_1, Z_2) = Z_1(\omega_i(Z_2)) - Z_2(\omega_i(Z_1)) - \omega_i([Z_1, Z_2]) = 0,$$

for all pairs of vector fields $Z_1, Z_2 \in D(P)$ because of complete integrability P, $[Z_1, Z_2] \in D(P)$.

Theorem 3.4 *Let* $\mathfrak{g} = \langle \overline{X}_1, \ldots, \overline{X}_m \rangle$ *be a transversal commutative Lie algebra of shuffling symmetries, and let* $\langle \omega_1, .., \omega_m \rangle$ *be a normalized basis in* Ann(P). *Then, all differential 1-forms* ω_i *are closed:*

$$d\omega_i = 0,$$

for $i = 1, \ldots, m.$

Corollary 3.1 *Under conditions of the above theorem and $H^1(M, \mathbb{R}) = 0$, the complete set of first integrals can be found by quadratures:*

$$H_1 = \int \omega_1, \ldots, H_m = \int \omega_m.$$

Example 3.4 Example (Distributions of Codimension 1) Let $P = \ker \omega$ be a completely integrable distribution of codimension one, and let X be a transversal symmetry of P, i.e. $\omega(X) \neq 0$. Then, the differential 1-form

$$\frac{\omega}{\omega(X)}$$

is closed, and

$$\int \frac{\omega}{\omega(X)}$$

is a first integral.

Example 3.5 Let $M = \mathbb{R} \times \mathbb{R} \times (0, 2\pi)$, with coordinates $x \in \mathbb{R}$, $y \in \mathbb{R}$, $\phi \in (0, 2\pi)$. The 1-form

$$\omega = 2 \sin^2 \frac{\phi}{2} \, dx + \sin \phi \, dy - y \, d\phi$$

defines the so-called oricycle distribution, and the vector field

$$X = \frac{\partial}{\partial x}$$

is a shuffling symmetry. The 1-form

$$\overline{\omega} = \frac{\omega}{\omega(X)} = dx + \frac{\sin \phi}{1 - \cos \phi} dy - \frac{y}{1 - \cos \phi} d\phi$$

is closed, and the function

$$H = \int \overline{\omega} = x + y \cot \frac{\phi}{2}$$

is a first integral.

Example 3.6 (Liouville–Arnold) Let $\left(M^{2n}, \Omega\right)$ be $2n$-dimensional symplectic manifold with structure 2-form Ω, and let the functions $(H_1, .., H_n)$ be independent and in involution, i.e.

$$dH_1 \wedge \cdots \wedge dH_n \neq 0,$$

and the Poisson brackets vanish: $[H_i, H_j] = 0$ for $i, j = 1, \ldots, n$.

Then, the level surfaces

$$L^c = \{H_1 = c_1, \ldots, H_n = c_n\} \subset M^{2n}$$

are Lagrangian submanifolds, and the Hamiltonian vector fields $X_{H_i}, i = 1, \ldots, n$ are tangent to L^c.

These vector fields are independent, and the involutivity conditions mean that they commute. We have $\dim L^c = n$ and, as we have seen, there exist such differential 1-forms $\omega_1^c, \ldots, \omega_n^c$ that $\omega_i^c (X_{H_j}) = \delta_{ij}$, which are closed. Then, integrals $F_i^c = \int_{L^c} \omega_i^c$ give us (multivalued) functions on L^c with linearly independent differentials. The submanifolds L^c give us a foliation of M, and we define functions $F_i, i = 1, \ldots, n$ on M by the requirement that their restrictions on L^c equal F_i^c.

Differentials of the functions $H_1, \ldots, H_n, F_1, \ldots, F_n$ are linearly Independent, and therefore, in simply connected domains, they are coordinates.

We have $[H_i, H_j] = 0$, and $[H_i, F_j] = \delta_{ij}$, and the flows along Hamiltonian vector fields X_{H_i} in these coordinates take the form

$$\dot{H}_j = 0, \ \dot{F}_j = \delta_{ij}.$$

Therefore, the equations for flows of Hamiltonian vector fields X_{H_i} are integrated in quadratures.

3.5.2 Symmetry Reduction

Let P be a completely integrable distribution, $\mathfrak{g} \subset \text{Sym}(P)$ a Lie algebra of shuffling symmetries, and $\mathfrak{j} \subset \mathfrak{g}$ an ideal in the Lie algebra. For any point $a \in M$, define a subspace $P_{\mathfrak{j}}(a) \subset T_a M$ formed by the space $P(a)$ and the space generated by values of vector fields from $\mathfrak{j}$ at the point. Assume that $\dim P_{\mathfrak{j}}(a)$ is constant, so that $P_{\mathfrak{j}} \supset P$ is a distribution. Then, the following result is valid.

Theorem 3.5

1. *The distribution $P_{\mathfrak{j}} \supset P$ is completely integrable, and*
2. *the quotient Lie algebra $\mathfrak{g}/\mathfrak{j}$ is a shuffling symmetry Lie algebra of the distribution $P_{\mathfrak{j}}$.*

Proof We have

- $[D(P), D(P)] \subset D(P)$, because P is completely integrable.
- $[\mathfrak{j}, D(P)] \subset D(P)$, because $\mathfrak{j}$ is symmetry ideal.
- $[\mathfrak{j}, \mathfrak{j}] \subset \mathfrak{j}$, because $\mathfrak{j}$ is an ideal.

- Therefore, $[D(P_j), D(P_j)] \subset D(P_j)$ and P_j is completely integrable.
- $[\mathfrak{g}, D(P)] \subset D(P)$ and $[\mathfrak{g}, \mathfrak{j}] \subset \mathfrak{j}$. Therefore, $\mathfrak{g}$ is a symmetry algebra of P_j, and $\mathfrak{g}/\mathfrak{j}$ is a shuffling symmetry algebra.

$\square$

Remark 3.2 This theorem shows that integration of P with symmetry algebra $\mathfrak{g}$ could be done in two steps:

1. Integration of the completely integrable distribution with symmetry algebra $\mathfrak{g}/\mathfrak{j}$.
2. Integration of the restrictions of distribution P on integral manifolds of distribution P_j by symmetry algebra $\mathfrak{j}$.

Assume now that algebra $\mathfrak{g}/\mathfrak{j}$ is commutative. Then, the first step could be done by quadratures due to theorem (3.4). The next step involves integration of distributions with symmetry algebra $\mathfrak{j}$, and if this algebra possesses an ideal $\mathfrak{j}_2 \subset \mathfrak{j}_1 = \mathfrak{j}$, such that $\mathfrak{j}_1/\mathfrak{j}_2$ is commutative, we could reduce it by quadratures. There is a special class of Lie algebras that can be exhausted by this procedure.

Let $\mathfrak{g}$ be a Lie algebra. A Lie algebra $\mathfrak{g}$ is said to be *solvable* if there is chain of subalgebras $\mathfrak{j}_i$

$$\mathfrak{g} \supset \mathfrak{j}_1 \supset \cdots \supset \mathfrak{j}_i \supset \mathfrak{j}_{i+1} \supset \cdots \supset \mathfrak{j}_k = 0,$$

such that $\mathfrak{j}_{i+1}$ is an ideal in $\mathfrak{j}_i$ and the quotient Lie algebra $\mathfrak{j}_i/\mathfrak{j}_{i+1}$ is commutative for every i.

A more constructive, but equivalent, definition uses the chain of *derived subalgebras*. Namely, the *derived Lie algebra* $\mathfrak{g}^{(1)} = [\mathfrak{g}, \mathfrak{g}]$ is the subalgebra of $\mathfrak{g}$ that consists of all linear combinations of Lie brackets of pairs of elements of $\mathfrak{g}$, and *derived series of the Lie algebra* is given by $\mathfrak{g}^{(i)} = [\mathfrak{g}^{(i-1)}, \mathfrak{g}^{(i-1)}]$, for $i = 1, 2, \ldots$.

We have chain of Lie subalgebras

$$\mathfrak{g} \supset \mathfrak{g}^{(1)} \supset \cdots \supset \mathfrak{g}^{(i)} \supset \mathfrak{g}^{(i+1)} \supset \cdots,$$

with commutative Lie algebras $\mathfrak{g}^{(i)}/\mathfrak{g}^{(i+1)}$, and the Lie algebra $\mathfrak{g}$ is *solvable* if $\mathfrak{g}^{(k)} = 0$, for some k.

Assume now that $\mathfrak{g}$ is a solvable Lie algebra of shuffling symmetries which is transversal to the completely integrable distribution P. Let $\mathfrak{g}^{(1)}$ be the first derived subalgebra and let $l = \text{codim}_\mathfrak{g} \mathfrak{g}^{(1)} > 0$. We choose a basis $X_1, \ldots, X_l, \ldots, X_m$ in the Lie algebra $\mathfrak{g}$ in such a way that $X_1, \ldots, X_l \notin \mathfrak{g}^{(1)}$, but $X_i \in \mathfrak{g}^{(1)}$, for $i \geqslant l+1$. We also choose a basis $\omega_1, \ldots, \omega_m$ in $\text{Ann}(P)$ such that $\omega_i(X_j) = \delta_{ij}$. Then,

$$d\omega_i(X_s, X_t) = -\omega_i([X_s, X_t]) = 0$$

for all $i = 1, \ldots, l$ and $s, t = 1, \ldots, m$.

Therefore, the differential 1-forms $\omega_i, i \leqslant l$, are closed and

$$H_i = \int \omega_i$$

are (in general multivalued) first integrals of the distribution P.

Moreover, the submanifolds

$$M_c = H_1^{-1}(c_1) \cap \cdots \cap H_l^{-1}(c_l)$$

are $\mathfrak{g}^{(1)}$-invariant because

$$X_i(H_j) = dH_j(X_i) = \omega_j(X_i) = 0,$$

if $j \leqslant l, i \geqslant l+1$.

Let P_c be the restriction of distribution P on the submanifold M_c. Then, P_c is a completely integrable distribution of the same dimension dim P and codimension dim $\mathfrak{g}^{(1)}$. Applying the above procedure in series to derived subalgebras $\mathfrak{g}^{(i)}$, we find the complete sequence of first integrals by integration of closed differential 1-forms, and the integral manifolds of P are given by quadratures: $H_1^{-1}(c_1) \cap \cdots \cap H_m^{-1}(c_m)$.

Theorem 3.6 (Lie–Bianchi) *Let P be a completely integrable distribution, and let $\mathfrak{g}$ be a solvable symmetry Lie algebra transversal to P, dim $\mathfrak{g}$ = codim P. Then, the distribution P is integrable by quadratures.*

3.5.3   Quadratures and Model Equations

In this section, we consider in more detail the case when the symmetry Lie algebra $\mathfrak{g} \subset \mathrm{Sym}\,(P)$ of a completely integrable distribution P acts in a transitive way on the set Sol$\,(P)$ of all maximal integral manifolds of P. In general, this set has very complicated structure, which is why we restrict ourselves to only consider the set Sol$_{\mathrm{loc}}\,(P)$ of local maximal integral submanifolds or (better to say) germs. In a small neighborhood of a point $a \in M$, this set is an open domain in $\mathbb{R}^m$, where $m = $ codim P.

So, we assume that dim $\mathfrak{g} \geqslant m$, and the value maps $\xi_a : \mathfrak{g} \to T_a M / P_a$ are surjections. The case when ξ_a are isomorphisms and $\mathfrak{g}$ is a solvable Lie algebra is completely covered by the Lie–Bianchi theorem.

To proceed with the general case, we take a homogeneous space G/H of a simply connected Lie group G, where Lie$(G) = \mathfrak{g}$ and Lie$(H) = \ker \xi_a$. The left action of the Lie group G on the homogeneous space G/H gives us the embedding λ of the Lie algebra $\mathfrak{g}$ into Lie algebra Vect$\,(G/H)$ of vector fields on G/H.

Let us consider a distribution $\widehat{P}$ on $M \times G/H$ generated at point $(a, b) \in M \times G/H$ by vectors of P_a and vectors $X_a + \lambda(X)_b$, where $X \in \mathfrak{g}$. In other words, the module $D(\widehat{P})$ is generated by $D(P)$ and vector fields of the form $X + \lambda(X)$, where $X \in \mathfrak{g}$.

This distribution is completely integrable, because $[D(P), D(P)] \subset D(P)$ and $[\mathfrak{g}, D(P)] \subset D(P)$. Moreover, codim $\widehat{P} = \dim G/H$ and maximal integral manifolds of $\widehat{P}$ are graphs of some maps $h : M \to G/H$, which we call *integral*.

Assume that we have an integral map h. Then, the tangent space to the graph at a point $(a, h(a))$ equals to $\widehat{P}_{(a,h(a))}$, the image of differential $h_{*,a}$ at a point $a \in M$, $\operatorname{Im} h_{*,a} = T_{h(a)}(G/H)$, and therefore $h_{*,a}$ is a surjection. Consider a submanifold $M_b = h^{-1}(b) \subset M$, $b \in G/H$. Then, codim $M_b = \dim G/H =$ codim P and $TM_b \subset P$. Therefore, M_b are maximal integral manifolds of P and to find them, if the integral map h is known, we should solve *functional equations* $h(x) = b$.

To construct an integral map h, we will use the following lifting method. Assume that the value of h at a point $a_0 \in M$ is fixed, $h(a_0) = b_0$, and M is a connected manifold. Then, to find value $h(a_1)$ at a point $a_1 \in M$, we take a path $\alpha(t)$, with $\alpha(t_0) = a_0$, $\alpha(t_1) = a_1$, and lift it to a path $\overline{\alpha}(t)$ on $M \times G/H$ in such a way that the curve $\overline{\alpha}(t)$ is an integral curve for the distribution $\widehat{P}$. Assume that the tangents $\dot{\alpha}(t)$ to the curve do not lie in the distribution P. Then, we can present them as linear combinations of values of vector fields in $\mathfrak{g}$, say

$$\dot{\alpha}(t) = q_1(t) \, X_{1,\alpha(t)} + \cdots + q_k(t) \, X_{k,\alpha(t)}, \tag{3.23}$$

for some functions $q_1(t), \ldots, q_k(t)$, where $X_1, \ldots, X_k$ is a basis in the Lie algebra $\mathfrak{g}$.

The path $\overline{\alpha}(t)$ on $M \times G/H$ is a lift of the path $\alpha(t)$ on M, and integral curve for distribution $\widehat{P}$, if and only if $\overline{\alpha}(t)$ satisfies the following equation, similar to (3.23):

$$\dot{\overline{\alpha}}(t) = q_1(t) \, \lambda(X_1)_{\overline{\alpha}(t)} + \cdots + q_k(t) \, \lambda(X_k)_{\overline{\alpha}(t)}. \tag{3.24}$$

In other words, in order to lift the path $\alpha(t)$ to the path $\overline{\alpha}(t)$, we have to find integral curves of a vector field of the form

$$\frac{\partial}{\partial t} + \sum_{i=1}^{k} q_i(t) \, \lambda(X_i) \tag{3.25}$$

on $\mathbb{R} \times G/H$, which correspond to paths $X(t) = \sum_{i=1}^{k} q_i(t) \, X_i$ on the Lie algebra $\mathfrak{g}$. We call equations of form (3.24) *model differential equations*.

Summarizing, we get the following generalization of the Lie–Bianchi theorem.

Theorem 3.7 *Let $\mathfrak{g}$ be a symmetry Lie algebra of a completely integrable distribution P on a connected manifold M such that the value maps $\xi_a : \mathfrak{g} \to T_a M / P_a$ are surjective. Then, integration of the distribution can be done by using solutions of model differential equations, corresponding to the Lie algebra symmetry, and solutions of a number of functional equations.*

We begin with the main properties of the model differential equations.

Theorem 3.8 *For any system of model ordinary differential equations associated with a path X_t on the Lie algebra $\mathfrak{g}$ and the vector field*

$$Z = \frac{\partial}{\partial t} + \lambda\,(X_t),$$

there is a path $g(t)$ on the group G with $g(0) = e$, such that any trajectory $y(t)$ of Z has the form

$$y(t) = g(t)y(0). \tag{3.26}$$

Moreover, this property defines the class of model equations completely.

Proof If the path $g(t)$ is given, then the path X_t is defined as follows:

$$X_t = g_*\,(t)^{-1}\left(\dot{g}\,(t)\right), \tag{3.27}$$

and from (3.26), we get $\dot{y}\,(t) = \lambda\,(X_t)$. On the other hand, if the path X_t is given, then the path $g(t)$ is found from equation (3.27). $\square$

Let us analyze one-dimensional models.

Theorem 3.9 (Lie) *Let a finite-dimensional Lie algebra $\mathfrak{g}$ act on $\mathbb{R}$ in a transitive way. Then, $\dim \mathfrak{g} \leqslant 3$, and locally the action is one of the following:*

1. $\dim \mathfrak{g} = 1$, $\mathfrak{g} = \langle \frac{\partial}{\partial x} \rangle$,
2. $\dim \mathfrak{g} = 2$, $\mathfrak{g} = \langle \frac{\partial}{\partial x}, x\frac{\partial}{\partial x} \rangle$, *and*
3. $\dim \mathfrak{g} = 3$, $\mathfrak{g} = \langle \frac{\partial}{\partial x}, x\frac{\partial}{\partial x}, x^2\frac{\partial}{\partial x} \rangle$.

A proof of this theorem can be found, for example, in [6]. The theorem shows that vector fields for model equations in the 1-dimensional case have the following forms:

1. $\frac{\partial}{\partial t} + a\,(t)\,\frac{\partial}{\partial x}$,
2. $\frac{\partial}{\partial t} + a\,(t)\,\frac{\partial}{\partial x} + b\,(t)\,x\frac{\partial}{\partial x}$, and
3. $\frac{\partial}{\partial t} + a\,(t)\,\frac{\partial}{\partial x} + b\,(t)\,x\frac{\partial}{\partial x} + c\,(t)\,x^2\frac{\partial}{\partial x}$.

The corresponding model equations are

1. $\dot{x}\,(t) = a\,(t)$,
2. $\dot{x}\,(t) = a\,(t) + b\,(t)\,x\,(t)$, and
3. $\dot{x}\,(t) = a\,(t) + b\,(t)\,x\,(t) + c\,(t)\,x\,(t)^2$.

Let functions $a(t)$ and $b(t)$ be given. To find solutions of model equations of types 1 and 2, we should add to our set of functions

1. the function $\int a(t)dt$ and
2. the functions $\exp\left(\int b(t)dt\right)$ and $\int a(t)\exp\left(-\int b(t)dt\right)dt$,

respectively.

This observation explains the idea of Liouville (1833) to introduce the field of elementary functions as the result of a series of Liouvillian extensions of the field of rational functions, i.e. extension by adding integrals and exponents of integrals, in other words, by adding solutions of model equations of the first and second types.

Model equations of the type 3 are known as Riccati equations and, as we will see later, they are related to linear ordinary equations of the second order as well as projective structures on the line.

More information on model equations corresponding to simple Lie groups can be found in [6].

3.5.4 The Lie Superposition Principle

Here, we use Theorem 3.8 to get more information about solutions of model equations.

As we have seen, finding solution $y(t)$ of the model equation is equivalent to finding a path $g(t)$ on the Lie group G, and $y(t) = g(t)y(0)$. Assume that we know k solutions, say $y_1(t), \ldots, y_k(t)$, then $y_i(t) = g(t)y_i(0)$, for all $i = 1, \ldots, k$.

From a geometrical point of view, this means that points $(y_1(t), \ldots, y_k(t)) \in (G/H)^k = (G/H) \times \cdots \times (G/H)$ and $(y_1(0), \ldots, y_k(0)) \in (G/H)^k$ can be transformed by a single transformation $g(t) \in G$, for any $k = 1, 2, \ldots$.

To analyze this situation, consider diagonal G-actions on direct products $(G/H)^k$, $g : a = (a_1, \ldots, a_k) \rightarrow ga = (ga_1, \ldots, ga_k)$. Then, the stationary group G_a is the intersection $G_{a_1} \cap \cdots \cap G_{a_k}$ of stationary subgroups G_{a_i} of points $a_i \in G/H$.

By the *stiffness* of a homogeneous manifold, we mean a number k such that the stationary groups G_a for general points $a = (a_1, \ldots, a_k) \in (G/H)^k$ are trivial. We call such points *regular*. Given two regular points $a, b \in (G/H)^k$, there is a unique element $\gamma(a, b) \in G$ such that $a = \gamma(a, b)b$.

A set of solutions $(y_1(t), \ldots, y_k(t))$ is said to be fundamental solution of the model equation if k is the stiffness of G/H and $(y_1(0), \ldots, y_k(0))$ is a regular point. Then, we define the path $g(t)$ as

$$g(t) = \gamma((y_1(t), \ldots, y_k(t)), (y_1(0), \ldots, y_k(0))),$$

and all solutions of the model equation have the form

$$y(t) = \gamma((y_1(t), \ldots, y_k(t)), (y_1(0), \ldots, y_k(0)))\, y(0). \tag{3.28}$$

The last formula is called the *Lie superposition principle*.

Example 3.7 Example (1D model equations)

1. Consider the case $G/H = \mathbb{R}$, $\mathfrak{g} = \langle \frac{\partial}{\partial x} \rangle$. Then, $\gamma (a, b) = a - b$. The fundamental solution is a solution of equation $\dot{y}(t) = A(t)$, and the superposition principle says that all solutions of the equation $\dot{x}(t) = A(t)$ have the form

$$x(t) = x(0) + (y(t) - y(0)) .$$

2. Let $G/H = \mathbb{R}$, $\mathfrak{g} = \langle \frac{\partial}{\partial x}, x \frac{\partial}{\partial x} \rangle$ and G be the group of all affine transformations of the line. The stiffness is $k = 2$ and

$$\gamma ((a_1, a_2), (b_1, b_2))\ x = \frac{(x - b_2)\, a_1 - (x - b_1)\, a_2}{b_1 - b_2} .$$

The fundamental solution of the model equation $\dot{x}(t) = A(t) + B(t)x(t)$ is pair of solutions $y_1(t)$, $y_2(t)$ such that $y_1(0) \neq y_2(0)$, and the superposition principle says that the general solution has the form

$$x(t) = \frac{(x(0) - y_2(0))\, y_1(t) - (x(0) - y_1(0))\, y_2(t)}{y_1(0) - y_2(0)} .$$

3. Consider $G/H = \mathbb{R}\mathbf{P}^1$, $\mathfrak{g} = \langle \frac{\partial}{\partial x}, x \frac{\partial}{\partial x}, x^2 \frac{\partial}{\partial x} \rangle$ with the group $G = \mathbf{SL}_2(\mathbb{R})$ of projective transformations

$$A : x \to \frac{a_{11}x + a_{12}}{a_{21}x + a_{22}},$$

where $A = \|a_{ij}\| \in \mathbf{SL}_2(\mathbb{R})$. It is known that any projective transformation of the projective line is completely determined by images of three distinct points. Therefore, the stiffness of $\mathbb{R}\mathbf{P}^1$ equals 3. It is also known that the cross-ratio

$$\frac{x - a_1}{x - a_2} \frac{a_3 - a_2}{a_3 - a_1}$$

is a projective invariant. Therefore, the element $\gamma (a, b) \in \mathbf{SL}_2(\mathbb{R})$ such that $y = \gamma ((x_1, x_2, x_3), (y_1, y_2, y_3))\, x$ can be found from the equation:

$$\frac{y - y_1}{y - y_2} \frac{y_3 - y_2}{y_3 - y_1} = \frac{x - x_1}{x - x_2} \frac{x_3 - x_2}{x_3 - x_1} .$$

Thus, a fundamental solution for the Riccati equation $\dot{x}(t) = A(t) + B(t)x(t) + C(t)x(t)^2$ is a triple of solutions $(y_1(t), y_2(t), y_3(t))$ with distinct initial values $y_1(0)$, $y_2(0)$, $y_3(0)$, and the general solution $y(t)$ can be found from the equation

$$\frac{y(0) - y_1(0)}{y(0) - y_2(0)} \frac{y_3(0) - y_2(0)}{y_3(0) - y_1(0)} = \frac{y(t) - y_1(t)}{y(t) - y_2(t)} \frac{y_3(t) - y_2(t)}{y_3(t) - y_1(t)} .$$

3.6 Ordinary Differential Equations

As we have seen above, the ordinary differential equation

$$F\left(x, y(x), y'(x), \ldots, y^{(k)}(x)\right) = 0 \qquad (3.29)$$

of order k, for functions $y(x)$ in one variable x, is represented as a submanifold $\mathcal{E}_F$ in the space of k-jets $\mathbf{J}^k$:

$$\mathcal{E}_F = \{F(x, u_0, u_1, \ldots, u_k) = 0\} \subset \mathbf{J}^k.$$

Its solutions $y(x)$ are represented by curves

$$L_y = \left\{u_0 = y(x), u_1 = y'(x), \ldots, u_k = y^{(k)}(x)\right\} \subset \mathcal{E}_F \subset \mathbf{J}^k.$$

These curves are integral for the restriction C_F of the Cartan distribution

$$C_k = \ker \omega_0 \cap \cdots \cap \ker \omega_{k-1}$$

on $\mathcal{E}_F$. Here,

$$\omega_i = du_i - u_{i+1} dx$$

are the *Cartan forms*.

As we have seen, the Cartan distribution C_k has dimension 2, and therefore the spaces of the distribution C_F are intersections

$$T\mathcal{E}_F \cap C_k = \ker dF \cap \ker \omega_0 \cap \cdots \cap \ker \omega_{k-1},$$

which are of dimension 2 or 1. The points $x_k \in \mathcal{E}_F$ where the intersection has dimension 2 we call *singular* and the points where the dimension equals 1 we call *regular*.

Since we have

$$dF = D_k(F)\, dx + \frac{\partial F}{\partial u_k} du_k \quad \mod C_k,$$

where

$$D_k = \frac{\partial}{\partial x} + u_1 \frac{\partial}{\partial u_0} + \cdots + u_k \frac{\partial}{\partial u_{k-1}},$$

points are singular if and only if

$$D_k (F) = 0, \quad \frac{\partial F}{\partial u_k} = 0. \tag{3.30}$$

At regular points, we have $\dim C_F = 1$ and the vector field

$$Z_F = \frac{\partial F}{\partial u_k} D_k - D_k (F) \frac{\partial}{\partial u_k} \tag{3.31}$$

is a basis in the distribution.

Thus, solutions L_y of (3.29) are trajectories of Z_F. However, there are trajectories of Z_F, which does not have the form L_y since the function x, in general, is not a coordinate on the trajectory. This situation appears every time when $\frac{\partial F}{\partial u_k} = 0$ on the trajectory.

Thus, we have two alternatives. Either we continue to consider solutions of (3.29) as smooth functions on the line, or we start to consider solutions as integral curves of distribution C_F. We shall follow the second alternative, and then we get solvability of our equation at regular points for free. Also, if $L \subset \mathcal{E}_F$ is an integral curve of C_F, then we can remove from L the points where $\frac{\partial F}{\partial u_k} = 0$. Then, we get

$$L \smallsetminus \left(\frac{\partial F}{\partial u_k} \right)^{-1} (0) = \bigcup_i L_i,$$

where each curve L_i has the form

$$L_i = L_{y_i(x)},$$

for some functions $y_i (x)$, each defined on its own interval. In this case, we call integral curves L *multivalued solutions* of (3.29).

There are also exceptional cases, when $\frac{\partial F}{\partial u_k} = 0$ at all points of L. Then, L has the form

$$L = \{x = \text{const}, u_0 = \text{const}_0, \ldots, u_{k-1} = \text{const}_{k-1}\},$$

where constants are chosen in such a way that

$$F (\text{const}, \text{const}_0, \ldots, \text{const}_{k-1}, u_k) = 0.$$

Example 3.8 Consider the Lissajous equation (see [7]):

$$\left(1 - x^2\right) y'' - xy' + \frac{a^2}{b^2} y = 0, \tag{3.32}$$

where $a \neq 0$ and $b \neq 0$ are some constants. Here,

$$F = \left(1 - x^2\right) u_2 - x u_1 + \frac{a^2}{b^2} u_0,$$

and

$$\mathcal{E}_F = \left\{ u_0 = \frac{\left(x^2 - 1\right) u_2 + x u_1}{a^2} b^2 \right\} \subset \mathbf{J}^2$$

is a smooth 3-dimensional submanifold in $\mathbf{J}^2$ with coordinates (x, u_1, u_2).

We have

$$\frac{\partial F}{\partial u_2} = 1 - x^2,$$

$$D_2 (F) = -3 x u_2 + \frac{a^2 - b^2}{b^2} u_1,$$

and therefore the two curves

$$x = \pm 1, \ u_1 = \pm \frac{a^2}{b^2} u_0, \ u_2 = \frac{a^2 \left(a^2 - b^2\right)}{3 b^4} u_0$$

consist of singular points of the equation.

Also, $\frac{\partial F}{\partial u_2} = 1 - x^2$, which implies that all integral curves that contain points $(x = \pm 1, u_0, u_1, u_2)$ are multivalued solutions. To find these solutions, we represent them in parametric forms

$$x = \cos (bt),$$

$$y = f (t),$$

for $-1 < x < 1$, and in the form

$$x = \pm \cosh (bt),$$

$$y = g (t),$$

for $x > 1$ or $x < -1$. Then, our equation in the first case takes the form

$$f'' + a^2 f = 0,$$

and the form

$$g'' - a^2 g = 0,$$

in the second case. Therefore, the solutions have the form

Fig. 3.1 The solution to (3.32) given by $x = \cos(5t)$ and $y = \sin(3t)$

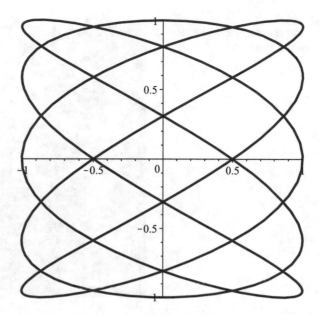

$$x = \cos(bt),$$

$$y = c_1 \cos(at) + c_2 \sin(at),$$

on intervals $|x| < 1$ and

$$x = \pm \cosh(bt),$$

$$y = c_1 \cosh(at) + c_2 \sinh(at),$$

on intervals $|x| > 1$. We plot the first type of solutions for $c_1 = 0$ and $c_2 = 1$ in the cases where $\frac{a}{b} = \frac{3}{5}$ and $\frac{a}{b} = \frac{3}{5\pi}$, respectively. The first picture (Fig. 3.1) gives exactly what we expect from multivalued solutions, but the second one (Fig. 3.2) is very far from the standard image.

3.7   ODE Symmetries

To simplify our exposition, we will assume that Eq. (3.29) is resolved with respect to the highest derivative and therefore has the form

$$\mathcal{E} = \{u_k = F(x, u_0, .., u_{k-1})\}. \tag{3.33}$$

Then, functions $(x, u_0, .., u_{k-1})$ are coordinates on $\mathcal{E}$, and the Cartan distribution $C_{\mathcal{E}}$ on $\mathcal{E}$ is given by forms

Fig. 3.2 The solution to (3.32) given by $x = \cos(5\pi t)$ and $y = \sin(3t)$

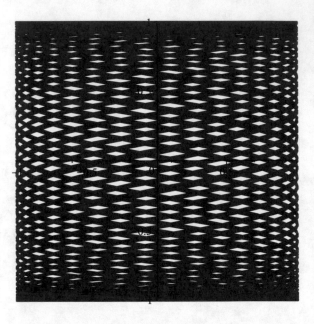

$$\omega_0 = du_0 - u_1 dx, \ldots, \omega_{k-2} = du_{k-2} - u_{k-1} dx, \; \omega_{k-1} = du_{k-1} - F dx$$

and generated by the truncated total derivative

$$D = \frac{\partial}{\partial x} + u_1 \frac{\partial}{\partial u_0} + \cdots + u_{k-1} \frac{\partial}{\partial u_{k-2}} + F \frac{\partial}{\partial u_{k-1}}.$$

Similarly to what we had in the general case, we have now the following expression for differentials of functions on $\mathcal{E}$ modulo the Cartan forms:

$$df = D(f) \, dx \quad \mathrm{mod} \; \langle \omega_0, .., \omega_{k-1} \rangle,$$

where $f = f(x, u_0, .., u_{k-1})$.

We have $\dim C_{\mathcal{E}} = 1$, and therefore this distribution is completely integrable, and the vector field D is obviously characteristic. We present shuffling symmetries in the form

$$X = a_0 \frac{\partial}{\partial u_0} + \cdots + a_{k-1} \frac{\partial}{\partial u_{k-1}},$$

where a_i are functions on $\mathcal{E}$.

This vector field is a symmetry of $C_{\mathcal{E}}$, or of $\mathcal{E}$, if

$$L_X(\omega_i) = 0 \quad \mathrm{mod} \; \langle \omega_0, \ldots, \omega_{k-1} \rangle,$$

for all $i = 0, \ldots, k-1$.

We have

$$L_X(\omega_i) = da_i - a_{i+1}dx = (D(a_i) - a_{i+1})dx \mod \langle \omega_0, \ldots, \omega_{k-1} \rangle, \quad (3.34)$$

for $i = 0, \ldots, k - 2$, and

$$L_X(\omega_{k-1}) = da_{k-1} - X(F)dx = (D(a_{k-1}) - X(F))dx \mod \langle \omega_0, \ldots, \omega_{k-1} \rangle. \quad (3.35)$$

Therefore, if we put $a_0 = \phi(x, u_0, .., u_{k-1})$, then the formulae (3.34) and (3.35) give us

$$a_i = D^i(\phi),$$

for $i = 0, \ldots, k - 1$ and

$$D^k(\phi) - X(F) = 0.$$

Therefore,

$$X = X_\phi = \phi \frac{\partial}{\partial u_0} + D(\phi) \frac{\partial}{\partial u_1} \cdots + D^{k-1}(\phi) \frac{\partial}{\partial u_{k-1}}, \quad (3.36)$$

and the last formula gives us condition on ϕ (Lie equation):

$$D^k(\phi) - \frac{\partial F}{\partial u_{k-1}} D^{k-1}(\phi) - \cdots - \frac{\partial F}{\partial u_1} D(\phi) - \frac{\partial F}{\partial u_0}\phi = 0. \quad (3.37)$$

Summarizing, we get the following result.

Theorem 3.10 *Let* Sym $(\mathcal{E})$ *be the Lie algebra of shuffling symmetries of ODE (3.33). Then, formula (3.36) gives the isomorphism of this Lie algebra with the space of smooth solutions to the Lie equation (3.37). The Lie algebra structure in* Sym $(\mathcal{E})$ *in terms of solutions (3.37), we call them generating functions, has the following form:*

$$[X_\phi, X_\psi] = X_{[\phi,\psi]},$$

where

$$[\phi, \psi] = X_\phi(\psi) - X_\psi(\phi). \quad (3.38)$$

Proof To prove (3.38), we remark that $\phi = X_\phi(u_0)$.
    Therefore,

$$[\phi, \psi] = X_{[\phi,\psi]}(u_0) = [X_\phi, X_\psi](u_0) = X_\phi(\psi) - X_\psi(\phi).$$

$\square$

Remark 3.3

1. If we compare formula (3.36) with description of symmetries of the general Cartan distribution, we see that when the generating function has the form $\phi = \phi(x, u_0, u_1)$, then the symmetries X_ϕ are restrictions of the contact symmetry of C_k on our differential equation. Moreover, symmetries with generating function ϕ, depending on $u_2, ..u_{k-1}$, are not anymore classical contact symmetries, and this can happen only when the order of the differential equation $k \geqslant 3$.
2. Contact symmetries with generating functions $\phi = b(x, u_0) + a(x, u_0) u_1$ are called point symmetries. They correspond to transformations given by vector fields

$$-a \frac{\partial}{\partial x} + b \frac{\partial}{\partial u_0}$$

on the plane $\mathbf{J}^0$ of the 0-jets (see [6] for more details).
3. Point symmetries, with generating functions $\phi = b(x)u_0 + a(x) u_1$, correspond to transformations, which are linear automorphisms of the line bundle $\pi_0 : \mathbf{J}^0 \to \mathbb{R}$. We call them linear symmetries.

3.7.1 Integration of ODEs with Commutative Symmetry Algebras

In this section, we will discuss an application of symmetries to integration of ODEs.

Assume that Eq. (3.33) has k linearly independent (in some domain) commuting symmetries $X_{\phi_0}, \ldots, X_{\phi_{k-1}}, [\phi_i, \phi_j] = 0$.

Remark that all symmetries X_{ϕ_i} preserve the Cartan distribution, and therefore

$$[X_{\phi_i}, D] = \lambda_i D,$$

for some functions λ_i, $i = 0, \ldots, k - 1$.

Applying both sides of this relation to the function x and using the relations $X_{\phi_i}(x) = 0$, $D(x) = 1$, we get $\lambda_i = 0$, i.e. $k + 1$ vector fields $D, X_{\phi_0}, \ldots, X_{\phi_{k-1}}$ commute and are linearly independent.

Consider differential forms

$$\omega_{-1} = dx, \omega_0, \ldots, \omega_{k-1}$$

on $\mathcal{E}$, then $\omega_{-1}(D) = 1$, $\omega_i(D) = 0$, for $i \geqslant 0$, and $\omega_i(X_{\phi_j}) = D^i(\phi_j)$, for $i, j \geqslant 0$.

Therefore, the matrix Ξ that we used in Sect. 3.5.1 has the form

$$
\varXi = \begin{Vmatrix} 1 & 0 & \cdots\cdots & 0 \\ 0 & \phi_1 & \cdots\cdots & \phi_{k-1} \\ \vdots & \vdots & \vdots & \vdots \\ \vdots & \vdots & \vdots & \vdots \\ 0 & D^{k-1}\phi_1 & \cdots\cdots & D^{k-1}(\phi_{k-1}) \end{Vmatrix} = \begin{Vmatrix} 1 & 0 \\ 0 & W \end{Vmatrix},
$$

where

$$
W = \begin{Vmatrix} \phi_1 & \cdots & \phi_{k-1} \\ \cdots & \cdots & \cdots \\ D^{k-1}\phi_1 & \cdots & D^{k-1}(\phi_{k-1}) \end{Vmatrix}.
$$

Therefore, the differential forms

$$
\varXi^{-1} \begin{Vmatrix} dx \\ \omega_0 \\ \vdots \\ \omega_{k-1} \end{Vmatrix} = \begin{Vmatrix} dx \\ W^{-1}\omega \end{Vmatrix} = \begin{Vmatrix} dx \\ \theta \end{Vmatrix},
$$

where

$$
\omega = \begin{Vmatrix} \omega_0 \\ \vdots \\ \omega_{k-1} \end{Vmatrix}, \qquad \theta = W^{-1}\omega, \tag{3.39}
$$

give us the dual coframe: $\theta_i\left(X_{\phi_i}\right) = \delta_{ij}$, and they are closed $d\theta_i = 0$.

Let

$$
f_i = \int \theta_i, \tag{3.40}
$$

then the functions $(x, f_0, \ldots, f_{k-1})$ are coordinates on $\mathcal{E}$ and since

$$
D\left(f_i\right) = \theta_i\left(D\right) = 0,
$$

they are first Integrals. In other words, solutions of $\mathcal{E}$ are given by relations

$$
f_0\left(x, u_0, \ldots, u_{k-1}\right) = \text{const}_0, \ldots, f_{k-1}\left(x, u_0, \ldots, u_{k-1}\right) = \text{const}_{k-1},
$$
$$\tag{3.41}$$

and the explicit dependency u_0 on x could be found by the elimination $u_1, \ldots, u_{k-1}$ from the above equations.

Theorem 3.11 *Solutions of ODE (3.33) of order k with k linearly independent commuting symmetries $X_{\phi_0}, \ldots, X_{\phi_{k-1}}$ can be found by quadrature (3.39, 3.40, 3.41).*

Example 3.9 Example ($k = 1$) Consider the first-order ODE

$$y' = F(x, y).$$

Its symmetries have the form

$$X_\phi = \phi(x, u_0) \frac{\partial}{\partial u_0},$$

where the function ϕ is a solution of the Lie equation

$$D(\phi) - F_{u_0}\phi = \phi_x + F\phi_{u_0} - F_{u_0}\phi = 0,$$

and

$$\theta = \frac{\omega_0}{\omega_0(X_\phi)} = \frac{du_0 - F dx}{\phi}.$$

If we consider the Lie equation as equation on F given ϕ, we get

$$F = \phi(x, u_0) \left(h(x) + \int \frac{\phi_x}{\phi^2} du_0 \right),$$

and

$$\theta = \frac{du_0}{\phi} - \left(h(x) + \int \frac{\phi_x}{\phi^2} du_0 \right) dx,$$

for arbitrary functions $h(x)$ and $\phi(x, u_0)$.

Example 3.10 Example ($k = 2$) Consider the second-order ODE

$$y'' = F(y, y').$$

Symmetries of this equation are vector fields of the form

$$X_\phi = \phi \frac{\partial}{\partial u_0} + \left(\phi_x + u_1\phi_{u_0} + F\phi_{u_1} \right) \frac{\partial}{\partial u_1},$$

where generating function $\phi = \phi(x, u_0, u_1)$ satisfies the Lie equation

$$D^2\phi - \frac{\partial F}{\partial u_1} D(\phi) - \frac{\partial F}{\partial u_0}\phi = 0,$$

with

$$D = \frac{\partial}{\partial x} + u_1 \frac{\partial}{\partial u_0} + F(u_0, u_1) \frac{\partial}{\partial u_1}.$$

Remark that the equation is invariant with respect to translations in x, and therefore $\phi = u_1$ is a solution of the Lie equation. In order to get a commutative algebra of symmetries, we assume that the generating function for the second symmetry is of the form $\phi(u_0)$. For this type of symmetry, the Lie equation takes the form

$$\phi_{u_0 u_0} u_1^2 - \phi F_{u_0} - \phi_{u_0}(u_1 F_{u_1} - F) = 0.$$

For example, equations with function F quadratic in u_1 have the form

$$F(u_0, u_1) = \frac{\phi_{u_0} + c_1}{\phi} u_1^2 + c_2 u_1 + c_3 \phi, \tag{3.42}$$

and $\phi = \phi(u_0)$ is the symmetry of the equation.

Let us take the representative with $c_1 = 0, c_2 = 0, c_3 = 1, \phi = \exp(u_0)$:

$$u_2 = u_1^2 + \exp(u_0).$$

It has commutative symmetry algebra ($\phi_1 = u_1, \phi_2 = \exp(u_0)$). Then,

$$W = \left\| \begin{matrix} \phi_1 & \phi_2 \\ D\phi_1 & D\phi_2 \end{matrix} \right\| = \left\| \begin{matrix} u_1 & \exp(u_0) \\ u_1^2 + \exp(u_0) & u_1 \exp(u_0) \end{matrix} \right\|,$$

and

$$\theta_1 = -u_1 \exp(-u_0) \omega_0 + \exp(-u_0) \omega_1$$
$$\theta_2 = \exp(-2u_0)\left(u_1^2 + u_0\right)\omega_0 - u_1 \exp(-2u_0) \omega_1$$

are closed 1-forms. We have

$$\theta_1 = dH_1, \qquad \theta_2 = dH_2,$$

where

$$H_1 = -x + \exp(-u_0) u_1 + c_1,$$
$$H_2 = -\frac{1}{2}u_1^2 \exp(-2u_0) - \exp(-u_0) + c_2,$$

and solutions of the equation are given by

$$y(x) = \ln\left(\frac{2}{2c_2 - c_1^2 - x^2 + 2c_1 x}\right),$$

where c_1 and c_2 are constants.

3.8 Schrödinger Type Equations

3.8.1 Actions of Diffeomorphisms on Schrödinger Type Equations

In this part, we consider linear second-order equations of the form

$$y'' + W(x)y = 0, \tag{3.43}$$

with the *potential* $W(x)$, and we study the cases when W is *integrable*, i.e. the Eq. (3.43) can be solved by quadratures. Because of linearity (3.43), we will restrict ourselves to linear symmetries $\phi = a(x)u_0 + b(x)u_1$.

Then, the Lie equation takes the form

$$D^2(\phi) + W(x)\phi = 0, \tag{3.44}$$

where

$$D = \frac{\partial}{\partial x} + u_1 \frac{\partial}{\partial u_0} - W(x)u_0 \frac{\partial}{\partial u_1},$$

or

$$\left(b'' + 2a'\right)u_1 + \left(a'' - 2Wb' - W'b\right)u_0 = 0$$

in explicit form. Therefore,

$$a = -\frac{1}{2}b' + c,$$

$$\phi = cu_0 + \phi_b,$$

$$\phi_b = bu_1 - \frac{b'}{2}u_0,$$

where the function $b(x)$ satisfies the following third-order differential equation:

$$b^{(3)} + 4Wb' + 2W'b = 0. \tag{3.45}$$

To understand the meaning of this equation, consider the action of point transformations on equations of type (3.43). If these transformations preserve the class of linear equations, they have to be of the following form:

$$T : (x, u_0) \to (y = Y(x), A(x)u_0).$$

Then, the image of (3.43) under this transformation is the following equation:

$$K_2 u_2 + K_1 u_1 + K_0 u_0 = 0,$$

where

$$K_2 = \frac{A}{(Y')^2},$$

$$K_1 = \frac{2A'Y' - AY''}{(Y')^3},$$

$$K_0 = \frac{-A'Y'' + Y'A'' - AW(Y)(Y')^3}{(Y')^3}.$$

Therefore, to get transformations preserving the class of Schrödinger type equations, we should require that $K_1 = 0$, and this requirement gives us the following class of transformations (up to a constant scaling of u_0):

$$T : (x, u_0) \to \left(Y(x), \sqrt{Y'}u_0\right), \quad Y' > 0. \tag{3.46}$$

Then, Eq. (3.43) will be transformed to the same type of equation with potential $\widetilde{W}$ equal to

$$\widetilde{W}(x) = (Y')^2 W(Y) + \frac{Y^{(3)}}{2Y'} - \frac{3(Y'')^2}{4(Y')^2}. \tag{3.47}$$

Notice that if we consider symmetric differential forms

$$g_W = W dy^2,$$

then the last equation takes the following form:

$$g_W = Y^*(g_W) - S(Y),$$

where

$$S(Y) = \left(-\frac{Y^{(3)}}{2Y'} + \frac{3(Y'')^2}{4(Y')^2}\right) dx^2$$

is the Schwarzian derivative.

It is easy to check that $S(Y) = 0$ if and only if Y is a projective transformation:

$$Y(x) = \frac{ax+b}{cx+d},$$

where

$$\left\| \begin{matrix} a & b \\ c & d \end{matrix} \right\| \in \mathbf{SL}_2(\mathbb{R}).$$

Therefore, the $\mathbf{SL}_2(\mathbb{R})$-action on the set of Schrödinger equations is equivalent to the $\mathbf{SL}_2(\mathbb{R})$-action on the space of quadratic differential forms.

3.8.2 Actions of the Diffeomorphism Group on Tensors

To understand the geometrical meaning of solutions of Schrödinger type equations, we reconsider actions of the diffeomorphism group on tensors.

Let M and N be manifolds, and let $\phi : M \to N$ be a diffeomorphism. We define action ϕ_* of ϕ on tensors in such a way that

$$(\psi \circ \phi)_* = \psi_* \circ \phi_*, \qquad (\phi_*)^{-1} = \left(\phi^{-1} \right)_*. \tag{3.48}$$

Functions   For $f \in C^\infty(M)$, we define $\phi_*(f) \in C^\infty(N)$ as follows:

$$\phi_*(f) = f \circ \phi^{-1}.$$

Remark that the difference of $\phi_*(f)$ from the more standard $\phi^*(h) = h \circ \phi$, where $h \in C^\infty(N)$ and $\phi^*(h) \in C^\infty(M)$, is the following:

- We define $\phi_*(f)$ for diffeomorphisms ϕ only but $\phi^*(h)$ defined for all smooth mappings ϕ.
- Since we have $(\phi \circ \psi)^* = \psi^* \circ \phi^*$, instead of (3.48), the correspondence $\phi \longmapsto \phi^*$ is not a group homomorphism, but $\phi \longmapsto \phi_*$ is.

Vector fields   Let $V \in \mathrm{Vect}(M)$ be a vector field on M, we define $\phi_*(V)$ by

$$\phi_*(V)(h) = \phi_*\left(V\left(\phi_*^{-1}(h) \right) \right),$$

where $h \in C^\infty(N)$ or, in operator form,

$$\phi_*(V) = \phi_* \circ V \circ \phi_*^{-1}.$$

Then, once more, we have property (3.48), and

$$\phi_* (fV) = \phi_* (f) \phi_* (V),$$

for all $f \in C^\infty (M)$.

Differential Forms Let $\omega \in \Omega^1 (M)$ be a differential 1-form on M, and then we define $\phi_* (\omega) \in \Omega^1 (N)$ as follows:

$$\langle \phi_* (\omega), Z \rangle = \phi_* \left\langle \omega, \phi_*^{-1} (Z) \right\rangle,$$

for all vector fields $Z \in \text{Vect}(N)$. For exterior differential forms of higher degree, we define

$$\phi_* (\omega_1 \wedge \cdots \wedge \omega_k) = \phi_* (\omega_1) \wedge \cdots \wedge \phi_* (\omega_k),$$

where $\omega_1 \in \Omega^1 (M), \ldots, \omega_k \in \Omega^1 (M)$ and $\omega_1 \wedge \cdots \wedge \omega_k \in \Omega^k (M)$. In a similar way, we define images of the symmetric differential forms and general tensors.

Once more, we have property (3.48) for maps of differential forms and tensors.

Coordinates Let $x = (x_1, .., x_n)$ be local coordinates on M and let $y = (y_1, .., y_n)$ be local coordinates on N. Assume that diffeomorphism ϕ has the following form in these coordinates:

$$\phi : x \to y = Y(x),$$
$$\phi^{-1} : y \to x = X(y).$$

Then,

$$f(x) \Longrightarrow \phi_* (f)(y) = f(X(y)),$$
$$h(y) \Longrightarrow \phi_*^{-1}(h)(x) = h(Y(x)).$$

Let $V = \frac{\partial}{\partial x_i} \in \text{Vect}(M)$. Then, $\phi_* (V) \in \text{Vect}(N)$ has the form

$$\phi_* (V) = \sum_{j=1}^{n} a_j (y) \frac{\partial}{\partial y_j},$$

and

$$a_j (y) = \phi_* (V)(y_j) = \phi_* \left(V \left(\phi_*^{-1} (y_j) \right) \right)$$

$$= \phi_* (V (Y_j(x))) = \phi_* \left(\frac{\partial Y_j}{\partial x_i} \right) = \frac{\partial Y_j}{\partial x_i} (X(y)).$$

Therefore,

$$\phi_* \left(\frac{\partial}{\partial x_i} \right) = \sum_{j=1}^{n} \frac{\partial Y_j}{\partial x_i} \left(X\left(y\right) \right) \frac{\partial}{\partial y_j}. \tag{3.49}$$

Let $\omega = dx_i \in \Omega^1 (M)$ be a differential 1-form. Then,

$$\phi_* \left(dx_i \right) = \sum_{j=1}^{n} b_j \left(y \right) \, dy_j,$$

and

$$b_j \left(y \right) = \phi_* \left(\left\langle dx_i, \phi_*^{-1} \left(\frac{\partial}{\partial y_j} \right) \right\rangle \right) = \frac{\partial X_i}{\partial y_j} \left(X\left(y\right) \right).$$

Therefore,

$$\phi_* \left(dx_i \right) = \sum_{j=1}^{n} \frac{\partial X_i}{\partial y_j} \left(X\left(y\right) \right) \, dy_j.$$

Solutions   As we have seen, the natural type of transformations for solutions of Eq. (3.43) is very similar to transformation of vector fields, with only one difference: instead of the multiplier $\frac{\partial Y}{\partial x}$ that we used for transformations of vector fields, we have to use multiplier $\sqrt{\frac{\partial Y}{\partial x}}$, i.e. solutions of the Schrödinger type equations behave like "$\frac{1}{2}-vector\ fields$." To check this hypothesis, we substitute square $b(x) = y(x)^2$ of a solution $y(x)$ of equation (3.43) in symmetry Eq. (3.45) and get zero.

Therefore, any product of solutions $b(x) = y_1(x)y_2(x)$ of the Schrödinger equation also satisfies equation (3.45). In other words, if $\langle y_1(x), y_2(x) \rangle$ is a fundamental set of solutions of Eq. (3.43), then $\langle y_1(x)^2, 2y_1(x)y_2(x), y_2^2(x) \rangle$ is a fundamental set of solutions of Eq. (3.45).

These functions represent vector fields

$$A = y_1(x)^2 \frac{\partial}{\partial x}, \ H = 2y_1\left(x\right)y_2(x)\frac{\partial}{\partial x}, \ B = y_2(x)^2 \frac{\partial}{\partial x}.$$

Assume that the Wronskian of y_1 and y_2 equals 1. Then,

$$[A, B] = H, \quad [H, A] = -2A, \quad [H, B] = 2B;$$

i.e. these vector fields satisfy the structure equations of the Lie algebra $\mathfrak{sl}(2, \mathbb{R})$.

To summarize, we have the following result.

Theorem 3.12

1. *The solution space of Eq. (3.45) is formed by pairwise products of solutions of Eq. (3.43), i.e. Eq. (3.45) is a symmetric square of Eq. (3.43).*
2. *The solution space of Eq. (3.45) is the $\mathfrak{sl}(2, \mathbb{R})$ Lie algebra with respect to bracket $[z_1, z_2] = z_1 z_2' - z_1' z_2$.*
3. *The Lie equation on symmetries of Schrödinger equation (3.43) defines a projective structure on the line: $\mathfrak{sl}(2, \mathbb{R}) \subset \mathrm{Vect}(\mathbb{R})$.*

Remark 3.4 By a projective structure on $\mathbb{R}$, we mean a covering $\mathbb{R}$ by intervals (U_α, t_α) with coordinates t_α such that on intersections $U_\alpha \cap U_\beta$, these coordinates are connected by projective transformations

$$t_\beta = \frac{a_{11}^{\beta\alpha} t_\alpha + a_{12}^{\beta\alpha}}{a_{21}^{\beta\alpha} t_\alpha + a_{22}^{\beta\alpha}}, \tag{3.50}$$

where

$$\left\| \begin{matrix} a_{11}^{\beta\alpha} & a_{12}^{\beta\alpha} \\ a_{21}^{\beta\alpha} & a_{22}^{\beta\alpha} \end{matrix} \right\| \in \mathbf{SL}_2(\mathbb{R}). \tag{3.51}$$

It is easy to check with formula (3.47) that locally any Schrödinger equation (3.43) could be transformed to the equation $y'' = 0$. The Lie algebra, corresponding to this equation, has the form

$$\mathfrak{sl}(2, \mathbb{R}) = \left\langle \frac{\partial}{\partial x}, x \frac{\partial}{\partial x}, x^2 \frac{\partial}{\partial x} \right\rangle. \tag{3.52}$$

In particular, any realization $\mathfrak{sl}(2, \mathbb{R}) \subset \mathrm{Vect}(\mathbb{R})$ is locally equivalent to model (3.52), and any two such realizations are connected by a projective transformation (3.50).

In other words, to define a projective structure on $\mathbb{R}$ is equivalent to have a representation of Lie algebra $\mathfrak{sl}(2, \mathbb{R})$ in Lie algebra of vector fields $\mathrm{Vect}(\mathbb{R})$, and it is also equivalent to have a Schrödinger equation (3.43).

Example 3.11 To the Schrödinger equation, $y'' + \omega^2 y = 0$ corresponds the Lie algebra

$$\mathfrak{sl}(2, \mathbb{R}) = \left\langle \sin^2(\omega x) \frac{\partial}{\partial x}, \sin(2\omega x) \frac{\partial}{\partial x}, \cos^2(\omega x) \frac{\partial}{\partial x} \right\rangle, \tag{3.53}$$

which is not equivalent to (3.52) on $\mathbb{R}$ because any nonvanishing vector field in (3.53) has an infinite number of zeroes, while those in (3.52) have not more than two.

3.8.3 Integration of Schrödinger Type Equations with Integrable Potentials

As we have seen, the Schrödinger type equations (3.43) have linear symmetries of the form $\phi_0 = u_0$ and $\phi_b = bu_1 - \frac{1}{2}b'u_0$, where the function $b = b(x)$ is a solution of the Lie equation

$$b^{(3)} + 4Wb' + 2W'b = 0.$$

They do commute $[\phi_0, \phi_b] = 0$, and therefore the Schrödinger equation can be integrated in quadratures if we know a nontrivial symmetry ϕ_b. In this case, we call potential W *integrable*.

Moreover, integrating the Lie equation with respect to W with given function b, we get the relation

$$W = \frac{c_b}{b^2} + \frac{1}{4}\left(\frac{b'}{b}\right)^2 - \frac{1}{2}\frac{b''}{b}, \qquad (3.54)$$

where c_b is a constant.

Remark that the solution space of the Lie equation is a Lie algebra, which is isomorphic to $\mathfrak{sl}_2$, and the constant c_b is proportional to the value of the Killing form on $\mathfrak{sl}_2$ on the vector $b \in \mathfrak{sl}_2$.

The relation (3.54) shows that if two potentials W and $\widetilde{W}$ have the same symmetry ϕ_b, then

$$\widetilde{W} - W = \frac{c}{b^2},$$

for some constant c.

This observation can also be used in the opposite way: if W is an integrable potential with symmetry ϕ_b, then the potential $W + \frac{c}{b^2}$ is also integrable with the same symmetry ϕ_b.

Example 3.12 The potentials

$$\frac{c}{\left(c_2 x^2 + 2c_1 x + c_0\right)^2},$$

$$\omega^2 + \frac{c}{\left(c_1 \sin\left(2\omega x\right) + c_2 \cos\left(2\omega x\right) + c_0\right)^2},$$

and

$$1 + \frac{1}{\left(\sin\left(2x\right) + 1\right)^2}$$

are integrable (Figs. 3.3 and 3.4).

Fig. 3.3 $w = \frac{1}{(x^2+1)^2}$

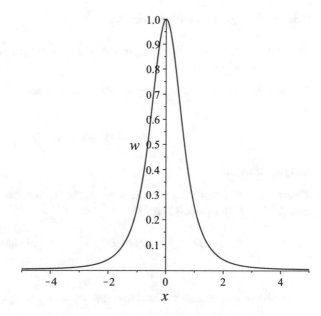

Fig. 3.4
$w = 1 + \frac{1}{(\sin^2(2x)+1)^2}$

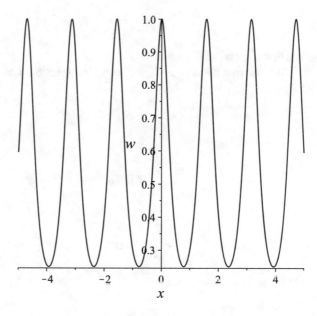

3.8.3.1 Integration by Symmetries

To integrate the Schrödinger equation with given symmetry ϕ_b, we will use the following observation:

Lemma 3.2 *Let ϕ and ψ be symmetries of the Schrödinger equation. Then, the function*

$$H = \phi D (\psi) - \psi D (\phi)$$

is a first integral.

Proof The conditions that ψ and ϕ are symmetries mean that $D^2 (\psi) + W\psi = 0$ and $D^2 (\phi) + W\phi = 0$. Therefore,

$$D (H) = D (\phi) D (\psi) - D (\psi) D (\phi) = 0.$$

$\square$

By applying this lemma to the symmetries $\psi = \phi_0 = u_0$ and $\phi = \phi_b$, we get that

$$H = \frac{c_b}{b}\phi_0^2 + \frac{1}{b}\phi_b^2$$

is a first integral of the equation.

Assume that $c_b = c_0^2 > 0$, and let y be a solution. Let

$$z = \frac{y}{\sqrt{|b|}},$$

then

$$H = c_0 z^2 + b^2 z'^2 = c^2.$$

Let

$$z = \frac{c}{c_0} \sin (\gamma), \qquad z' = \frac{c}{b} \cos (\gamma),$$

for some function γ. Then,

$$\gamma' = \frac{c_0}{b},$$

and therefore solutions y have the form

$$y = \frac{c}{c_0}\sqrt{|b|} \sin \left(c_0 \int \frac{dx}{b}\right).$$

In a similar way, we get

$$y = \frac{c}{c_0}\sqrt{|b|}\,\sinh\left(c_0 \int \frac{dx}{b}\right)$$

for the case $c_b = -c_0^2 < 0$, and

$$y = c\sqrt{|b|}\int \frac{dx}{b}$$

for the case $c_b = 0$.

Finally, we conclude that knowledge of a single nontrivial linear symmetry ϕ_b gives us

- Potential function

$$W = \frac{c_b}{b^2} + \frac{1}{4}\left(\frac{b'}{b}\right)^2 - \frac{1}{2}\frac{b''}{b}.$$

- Fundamental solution (if $c_b = c_0^2 > 0$)

$$y_1 = \sqrt{|b|}\,\sin\left(c_0 \int \frac{dx}{b}\right),\ y_2 = \sqrt{|b|}\,\cos\left(c_0 \int \frac{dx}{b}\right).$$

- Linear symmetries

$$b_1 = b,\, b_2 = b\sin\left(2c_0 \int \frac{dx}{b}\right),\, b_3 = b\cos\left(2c_0 \int \frac{dx}{b}\right).$$

- Integrable potentials

$$W + \sum_i \frac{k_i}{(c_{1i}b_1 + c_{2i}b_2 + c_{i3}b_3)^2},\quad \text{etc.}$$

Example 3.13 For the case

$$W = 1 + \frac{2}{(2 + \sin 2x)^2},\ b = ? + \sin 2x,\ c_b - 5,$$

we have (Fig. 3.5)

$$y_1 = \sqrt{2 + \sin 2x}\,\sin\left(\frac{2}{\sqrt{3}}\arctan\left(\frac{2\tan x + 1}{\sqrt{3}}\right)\right),$$

$$y_2 = \sqrt{2 + \sin 2x}\,\cos\left(\frac{2}{\sqrt{3}}\arctan\left(\frac{2\tan x + 1}{\sqrt{3}}\right)\right).$$

Fig. 3.5 $w = 1 + \frac{2}{(2+\sin 2x)^2}$

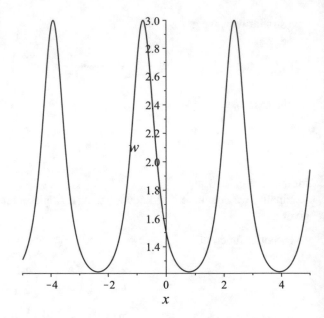

3.8.3.2 Lame Equation

Let us consider the case when the potential function W generates linear symmetry $\phi_W = W u_1 - \frac{1}{2} W' u_0$ for the corresponding Schrödinger equation.

In other words, the lift of the vector field

$$W(x) \frac{\partial}{\partial x}$$

in the bundle of $\frac{1}{2}$-vector fields is a symmetry of the Schrödinger equation.

Putting $b = W$ in the Lie equation gives us differential equation for potential:

$$W^{(3)} + 6W W' = 0,$$

or

$$W'' + 3W^2 + c_1 = 0$$

for some constant c_1. Therefore,

$$\frac{1}{2} W'^2 + W^3 + c_1 W + c_2 = 0,$$

for some constant c_2, and

$$W = -2\wp,$$

where $\wp$ is the Weierstrass p-function with invariants $g_2 = -c_1$ and $g_3 = c_2/2$.
Therefore, the Lame equation

$$y'' - 2\wp y = 0$$

has linear symmetry

$$\phi_\wp = \wp u_1 - \frac{1}{2}\wp' u_0,$$

with constant $c_\wp = -g_3$.

The fundamental solution (in the case, $g_3 < 0$) has the form

$$y_1 = \sqrt{|\wp|}\sin\left(\sqrt{-g_3}\int\frac{dx}{\wp}\right),\ y_2 = \sqrt{|\wp|}\cos\left(\sqrt{-g_3}\int\frac{dx}{\wp}\right),$$

and all linear symmetries are generated by

$$b_1 = \wp,\ b_2 = \wp\sin\left(2\sqrt{-g_3}\int\frac{dx}{b}\right),\ b_3 = \wp\cos\left(2\sqrt{-g_3}\int\frac{dx}{b}\right),$$

and the potentials

$$-2\wp + \sum_i \frac{k_i}{\wp^2\left(c_{1i} + c_{2i}\sin\left(2\sqrt{-g_3}\int\frac{dx}{b}\right) + c_{i3}\cos\left(2\sqrt{-g_3}\int\frac{dx}{b}\right)\right)^2}$$

are integrable.

3.8.3.3 Eigenvalue Problem

Here, we consider equations of the form

$$y'' + (W - \lambda)y = 0. \tag{3.55}$$

At first, we consider equations that have symmetries that are linear in λ:

$$b = b_0 + \lambda b_1.$$

Collecting terms with λ^2, λ, and λ^0, we get the following relations: $b_1 = $ const, $b_0 = -\frac{W}{2}$, and b_0 is a symmetry of W. Therefore, we get the discussed case of the Lame equation:

$$y'' - (\lambda + 2\wp)\, y = 0,$$

with symmetry

$$b = \lambda + \wp.$$

Then, the functions

$$y = C_1\sqrt{|\wp + \lambda|}\, \sin\left(\sqrt{-g_3}\int \frac{dx}{\wp + \lambda}\right) + C_2\sqrt{|\wp + \lambda|}\, \cos\left(\sqrt{-g_3}\int \frac{dx}{\wp + \lambda}\right)$$

satisfy equation (3.55).

Assume that we are looking for eigenvalues for the boundary value problem $y(a) = y(b) = 0$ on the interval $[a, b]$, where the function $\wp$ has no singularities.

Then, the boundary value problem gives us a linear system for the coefficients C_1 and C_2 of the solution

$$y(x) = C_1\sqrt{|\wp(x) + \lambda|}\, \sin\left(\sqrt{-g_3}\int_a^x \frac{dx}{\wp + \lambda}\right)$$
$$+ C_2\sqrt{|\wp(x) + \lambda|}\, \cos\left(\sqrt{-g_3}\int_a^x \frac{dx}{\wp + \lambda}\right).$$

We have

$$C_2\sqrt{|\wp(a) + \lambda|} = 0,$$

at the point $x = a$ and

$$C_1\sqrt{|\wp(b) + \lambda|}\, \sin\left(\sqrt{-g_3}\int_a^b \frac{dx}{\wp + \lambda}\right) + C_2\sqrt{|\wp(b) + \lambda|}\, \cos\left(\sqrt{-g_3}\int_a^b \frac{dx}{\wp + \lambda}\right) = 0$$

at the point $x = b$.

Therefore, solutions of the equation

$$\sqrt{|\wp(a) + \lambda|}\sqrt{|\wp(b) + \lambda|}\, \sin\left(\sqrt{-g_3}\int_a^b \frac{dx}{\wp + \lambda}\right) = 0$$

are eigenvalues for given boundary problem, or

$$\lambda = -\wp(a),\ \lambda = -\wp(b),$$

$$\int_a^b \frac{dx}{\wp + \lambda} = \frac{\pi k}{\sqrt{-g_3}}, k = 0, \pm 1, \pm 2, \ldots.$$

are the eigenvalues.

Similar results are valid for eigenvalue problems with symmetries that are polynomial in λ (see [8]).

Acknowledgments The research was partially supported by Grant RFBR No. 18-29-10013.

References

1. Alekseevskij, D.V., Vinogradov, A.M., Lychagin, V.V.: Basic ideas and concepts of differential geometry. Geometry I. Encyclopaedia Math. Sci., 28, Springer, Berlin (1991)
2. Bluman, G.W., Anco, S.C.: Symmetry and integration methods for differential equations. Applied Mathematical Sciences, 154. Springer-Verlag, New York (2002)
3. Bocharov, A.V., Chetverikov, V.N., Duzhin, S.V., Khor'kova, N.G., Krasil'shchik, I.S., Samokhin, A.V., Torkhov, Yu N., Verbovetsky, A.M., Vinogradov, A.M.: Symmetries and conservation laws for differential equations of mathematical physics. Translations of Mathematical Monographs, 182. American Mathematical Society, Providence, RI (1999)
4. Duzhin, S.V., Lychagin, V.V.: Symmetries of distributions and quadrature of ordinary differential equations. Acta Appl. Math. **24**(1), 29–57 (1991)
5. Krasil'shchik, J., Verbovetsky, A.: Geometry of jet spaces and integrable systems. J. Geom. Phys. **61**(9), 1633–1674 (2011)
6. Kushner, A., Lychagin, V., Rubtsov, V.: Contact geometry and non-linear differential equations. Encyclopedia of Mathematics and Its Applications, 101. Cambridge University Press, Cambridge (2007)
7. Kushner, A., Lychagin, V., Slovak, J.: Lectures on geometry of Monge-Ampere equations with Maple. In: Kycia, R., Ulan, M., Schneider, E. (eds.) Nonlinear PDEs, Their Geometry, and Applications., pp. 53–94. Birkhauser, Cham (2019)
8. Lychagin, V., Lychagina, O.: Finite dimensional dynamics for evolutionary equations. Nonlinear Dyn. **48**, 29–48 (2007)
9. Olver, P.J.: Applications of Lie groups to differential equations. Springer-Verlag (1993)
10. Shnider, S., Winternitz, P.: Classification of systems of nonlinear ordinary differential equations with superposition principles. J. Math. Phys. **25**(11), 3155–3165 (1984)
11. Krasil'shchik, I.S., Lychagin, V.V., Vinogradov, A.M.: Geometry of jet spaces and nonlinear partial differential equations. Advanced Studies in Contemporary Mathematics, 1. Gordon and Breach Science Publishers, New York (1986)

Chapter 4
Finite Dimensional Dynamics of Evolutionary Equations with Maple

Alexei G. Kushner and Ruslan I. Matviichuk

4.1 Introduction

The theory of finite dimensional dynamics is a natural development of the theory of dynamical systems. Dynamics make it possible to find families that depend on a finite number of parameters among all solutions of evolutionary differential equations.

The basic ideas and methods of this theory were formulated in [7, 13]. In the same papers, finite dynamics were constructed for the Kolmogorov–Petrovsky–Piskunov and the Korteweg–de Vries equations.

Second-order dynamics of the Burgers–Huxley equation were constructed in [10].

Dynamics of third order were found for equations of the Rapoport–Leas type arising in the theory of two-phase filtration. These dynamics were used for constructing attractors [2, 3].

The paper is devoted to the finite dimensional dynamics of some evolution equations that arise in physics, mathematical biology, and mathematical economics. Among them are the Fisher–Kolmogorov–Petrovsky–Piskunov [21] equation and its generalization and the Black–Scholes equation [4].

When finding dynamics, we have to carry out calculations in jet spaces. This leads to cumbersome formulas. To facilitate calculations and avoid mistakes, we use the packages DifferentialGeometry and JetCalculus of the system of symbolic

A. G. Kushner (✉)
Moscow Pedagogical State University, Moscow, Russia
e-mail: kushner@physics.msu.ru

R. I. Matviichuk
Lomonosov Moscow State University, Moscow, Russia

© The Author(s), under exclusive license to Springer Nature Switzerland AG 2021   123
M. Ulan, E. Schneider (eds.), *Differential Geometry, Differential Equations, and Mathematical Physics*, Tutorials, Schools, and Workshops in the Mathematical Sciences, https://doi.org/10.1007/978-3-030-63253-3_4

calculations Maple. A description of the basics of working with these packages can be found in [20].

Examples of dynamics calculations are considered and the program codes are given. These codes, with minor modifications, can be used to compute dynamics and find exact or approximate solutions to other evolutionary equations.

The structure of this paper is as follows.

In the first two sections, we give basic definitions and describe methods of the theory. The details can be found in [5, 7, 11, 13].

In the third section, we calculate first- and second-order dynamics of the Fisher–Kolmogorov–Petrovsky–Piskunov (FKPP) equation

$$u_t = u_{xx} + f(u)$$

and apply them to construct approximate solutions.

Note that in the considered example, the dynamics equation is solved exactly, but the group of shifts along the evolutionary vector field was found only approximately.

In the fourth section, we consider the reaction–diffusion equation with a convection term. This equation differs from the FKPP equation by the presence of a first-order derivative with respect to x (see [16, 18, 19]):

$$u_t + H(u)_x = u_{xx} + f(u).$$

It is proved that this equation has first-order dynamics for any smooth functions H and f, but second-order dynamics exist only when the function H is quadratic and the function f is cubic (see Theorem 4.3).

The fifth section is devoted to the Black–Scholes equation that came from mathematical finance theory. We construct two series of its exact solutions.

Some Maple files can be found on the website, d-omega.org.

4.2   Symmetries of ODEs

Any ordinary differential equation

$$y^{(k+1)} = h\left(x, y, y', y'', \ldots, y^{(k)}\right) \tag{4.1}$$

can be considered as a one-dimensional distribution P on the jet space $J^k(\mathbb{R})$. This distribution is generated by the vector field

$$\mathcal{D} = \frac{\partial}{\partial x} + y_1 \frac{\partial}{\partial y_0} + \cdots + y_k \frac{\partial}{\partial y_{k-1}} + h \frac{\partial}{\partial y_k}.$$

Here, $x, y_0, y_1, \ldots, y_k$ are canonical coordinates on $J^k(\mathbb{R})$ [9].

Integral curves of P are prolongations of trajectories of Eq. (4.1) into the space $J^k(\mathbb{R})$.

Definition 4.1 A vector field X on $J^k(\mathbb{R})$ is called an *infinitesimal symmetry* of Eq. (4.1) if translations along X preserve P.

All infinitesimal symmetries form the Lie algebra with respect to the Lie bracket. We denote this algebra by Symm P.

Definition 4.2 An infinitesimal symmetry is called *characteristic* if translations along it preserve each integral curve of the distribution P.

Characteristic symmetries form an ideal in Symm P, which we denote by Char P.

Definition 4.3 The quotient Lie algebra

$$\text{Shuff P} := \text{Symm P} / \text{Char P}$$

is called the Lie algebra of *shuffling* symmetries.

Each shuffling symmetry can be identified with a vector field of the form

$$S_\phi = \phi \frac{\partial}{\partial y_0} + \mathcal{D}(\phi) \frac{\partial}{\partial y_1} + \mathcal{D}^2(\phi) \frac{\partial}{\partial y_2} + \cdots + \mathcal{D}^k(\phi) \frac{\partial}{\partial y_k},$$

where ϕ is a function on $J^k(\mathbb{R})$ that is called a *generating function* of the corresponding shuffling symmetry.

If the function h does not depend on x, then y_1 is a generating function of a symmetry of Eq. (4.1).

4.3 Flows on ODE's Solution Spaces

Consider the following evolutionary partial differential equation:

$$\frac{\partial u}{\partial t} = \phi \left(x, u, \frac{\partial u}{\partial x}, \ldots, \frac{\partial^k u}{\partial x^k} \right). \tag{4.2}$$

Let $\phi = \phi(x, y_0, \ldots, y_k)$ be a generating function of some shuffling symmetry of Eq. (4.1), and let Ψ_t be the translation along the vector field S_ϕ from $t = 0$ to t. Let $L_{y(x)} = \{y_0 = y(x)\}$ be the graph of some solution $y = y(x)$ of Eq. (4.1), and let

$$L_{y(x)}^{(k)} = \{y_0 = y(x), y_1 = y'(x), \ldots, y_k = y^{(k)}(x)\} \tag{4.3}$$

be its prolongation into the space $J^k(\mathbb{R})$. Shifting the curve $L^{(k)}_{y(x)}$ along the trajectories of the vector field S_ϕ, we get the surface

$$\Phi_t\left(L^{(k)}_{y(x)}\right) \subset J^k(\mathbb{R}^2)$$

that is a prolongation of the graph of a solution of evolution equation (4.2). Here, $J^k(\mathbb{R}^2)$ is the k-jet space of functions with two independent variables t and x. Describe two methods for constructing solutions of equation (4.2).

Method 1 The space of solutions of equation (4.1) can be identified with the space $\mathbb{R}^{k+1}$ by indicating the initial data of solutions at a fixed point $x = x_0$. Then, the shift transformation Φ_t defines the transformation of the space $\mathbb{R}^{k+1}$ with coordinates $y_0, y_1, \ldots, y_k$. Therefore, we can consider transformations of this space instead of transforming curves. Such transformations are given by shifts $\overline{\Phi}_t$ along the vector field

$$E_\phi = \overline{\phi}\frac{\partial}{\partial y_0} + \mathcal{D}(\overline{\phi})\frac{\partial}{\partial y_1} + \mathcal{D}^2(\overline{\phi})\frac{\partial}{\partial y_2} + \cdots + \mathcal{D}^k(\overline{\phi})\frac{\partial}{\partial y_k},$$

where $\overline{\phi}$ is a restriction of the function ϕ to Eq. (4.1).

Let $y = y(x; \mathbf{a})$ be the solution of equation (4.1) with initial data

$$y(x_0) = a_0, \, y'(x_0) = a_1, \ldots, \, y^{(k)}(x_0) = a_k.$$

Applying the transformation $\overline{\Phi}_t$ to the point $\mathbf{a} = (a_0, \ldots, a_k)$, we obtain a one-parameter family $y(x; \overline{\Phi}_t(\mathbf{a}))$ of solutions of equation (4.1). Then, the function

$$u(t, x) = y(x, \overline{\Phi}_t(\mathbf{a}))$$

is a solution of the evolutionary Eq. (4.2) with the initial data $u(0, x) = y(x; \mathbf{a})$.

Method 2 The transformation Φ_t acting on the jet space $J^k(\mathbb{R})$ generates a transformation Φ_t^* acting on functions. Let Φ_t^{-1} be the inverse transformation for Φ_t. Curve (4.3) is generated by the system of equalities

$$y_0 - y(x) = 0, \, y_1 - y'(x) = 0, \ldots, \, y_k - y^{(k)}(x) = 0. \tag{4.4}$$

Applying the transformation $\left(\Phi_t^{-1}\right)^*$ to (4.4), we obtain the following systems:

$$\Psi^0(t, x, y_0, \ldots, y_k) = 0, \quad \Psi^1(t, x, y_0, \ldots, y_k) = 0, \ldots, \Psi^k(t, x, y_0, \ldots, y_k)=0.$$

Solving it with respect to $y_0, \ldots, y_k$, we find a coordinate representation of the curve $\Phi_t\left(L^{(k)}_{y(x)}\right)$:

$$y_0 = Y_0(t, x), \, y_1 = Y_1(t, x), \ldots, y_k = Y_k(t, x). \tag{4.5}$$

The function $u(t, x) = Y_0(t, x)$ is a solution of equation (4.2). The remaining functions in (4.5) correspond to the partial derivatives:

$$\frac{\partial^j u}{\partial x^j} = Y_j(t, x), \quad j = 1, \ldots, k.$$

The first method is convenient when the solution of equation (4.1) or the shift transformation Φ_t can be found only approximately. The second method, on the contrary, is applicable in the case when the solution and shift transformation can be found explicitly.

Definition 4.4 If ϕ is a generating function of a shuffling symmetry of Eq. (4.1), then Eq. (4.1) is called a (finite dimensional) *dynamics* of Eq. (4.2). The number $k + 1$ is called the *order* of the dynamics.

Thus, an evolutionary equation determines a flow on the solution space of an ordinary differential equation.

The following theorem (see [2]) provides a method for calculating finite dimensional dynamics of evolutionary equations.

Theorem 4.1 *The ordinary differential equation*

$$F = y_{k+1} - h(x, y_0, y_1, \ldots, y_k) = 0$$

is a dynamics of evolutionary equation (4.2) if and only if

$$[\phi, F] = 0 \bmod \mathbf{DF}, \tag{4.6}$$

where $\mathbf{DF} = \langle F, D(F), D^2(F), \ldots \rangle$ *is the differential ideal generated by the function F,*

$$D = \frac{\partial}{\partial x} + y_1 \frac{\partial}{\partial y_0} + y_2 \frac{\partial}{\partial y_1} + \cdots$$

is the operator of total derivative, and

$$[\phi, F] = \sum_{i \geq 0} \left(\frac{\partial \phi}{\partial y_i} D^i(F) - \frac{\partial F}{\partial y_i} D^i(\phi) \right)$$

is the Poisson–Lie bracket.

The Poisson–Lie bracket is a prolongation of the classical Poisson bracket into the jet space (see, for example, [11]). Note that the Poisson–Lie bracket is skew-symmetric $\mathbb{R}$-bilinear and satisfies the Jacobi identity. From the skew-symmetric property, it follows that $[\phi, \phi] = 0$, and therefore the equation $\phi = 0$ is a dynamics of Eq. (4.2).

4.4   The Fisher–Kolmogorov–Petrovsky–Piskunov Equation

The equation

$$u_t = u_{xx} + f(u) \tag{4.7}$$

is known as the Fisher–Kolmogorov–Petrovsky–Piskunov equation or the reaction–diffusion equation.

It describes processes of heat and mass transfer, propagation of a dominant gene [6, 8], propagation of flame [21], reaction–diffusion [14], and ferroelectric domain wall motion in an electric field [17]. For example, Eq. (4.7) with

$$f(u) = (1 - u^2)(m - u),$$

$-1 < m \leqslant 0$, describes active transmission of an electric impulse in neuron, and it is known as Nagumo's equation [15].

B. Kruglikov and O. Lychagina [7] presented an analysis of finite dimensional dynamics of Eq. (4.7).

4.4.1   Second-Order Dynamics

Equation (4.7) admits second-order dynamics if the function $f(u)$ is cubic (see [7]):

$$f(u) = f_3 u^3 + f_2 u^2 + f_1 u + f_0,$$

where $f_0, \ldots, f_3 \in \mathbb{R}$. Then,

$$\phi = y_2 + f_3 y_0^3 + f_2 y_0^2 + f_1 y_0 + f_0.$$

Find second-order dynamics in the form of the Liénard equation [12], i.e. put

$$F := y_2 - A(y_0)y_1 - B(y_0), \tag{4.8}$$

where A and B are some smooth functions. Consider two cases.

Case 1: $f_3 > 0$ Then, we can put $f_3 = 2q^2$. The restriction of the Poisson–Lie bracket to dynamics (4.8) gives us the following system of equations:

$$\begin{cases} fB' - Bf' = 0, \\ (2B - f)A' = 0, \\ A'' = 0, \\ B'' + 2AA' + 12q^2 y_0 + 2f_2 = 0. \end{cases}$$

Solving this system, we get

$$A(y_0) = A_1 y_0 + A_0$$

and

$$B(y_0) = -\frac{1}{3}(A_1^2 + 6q^2)y_0^3 - (A_0 A_1 + f_2)y_0^2 + B_1 y_0 + B_0,$$

where A_0, A_1, B_0, and B_1 are constants that we find from the first two equations: A_0 is arbitrary, and

$$A_1 = 0, \quad B_0 = -f_0, \quad B_1 = -f_1.$$

Then, dynamics (4.8) has the form

$$F = y_2 + 2q^2 y_0^3 + f_2 y_0^2 - A_0 y_1 + f_1 y_0 + f_0 = \phi - A_0 y_1. \tag{4.9}$$

Therefore, the restriction ϕ to equation $F = 0$ is

$$\overline{\phi} = A_0 y_1.$$

Case 2: $f_3 < 0$ We can put $f_3 = -2q^2$, and we get the equation

$$B'' + 2AA' - 12q^2 y_0 + 2f_2 = 0$$

instead of the last equation in system (4.4.1). Then, $A(y_0) = A_1 y_0 + A_0$ and

$$B(y_0) = -\frac{1}{3}(A_1^2 - 6q^2)y_0^3 - (A_0 A_1 + f_2)y_0^2 + B_1 y_0 + B_0,$$

and we get three solutions:

1. A_0 is arbitrary, $A_1 = 0$, $B_0 = -f_0$, $B_1 = -f_1$;
2. $A_0 = -\dfrac{f_2}{2q}$, $A_1 = 3q$, $B_0 = \dfrac{f_0}{2}$, $B_1 = \dfrac{f_1}{2}$;
3. $A_0 = \dfrac{f_2}{2q}$, $A_1 = -3q$, $B_0 = \dfrac{f_0}{2}$, $B_1 = \dfrac{f_1}{2}$.

So, we get the following dynamics:

$$F_1 = y_2 - A_0 y_1 - 2q^2 y_0^3 + f_2 y_0^2 + f_1 y_0 + f_0, \tag{4.10}$$

$$F_2 = y_2 - \left(3q y_0 - \frac{f_2}{2q}\right) y_1 + q^2 y_0^3 - \frac{f_2}{2} y_0^2 - \frac{f_1}{2} y_0 - \frac{f_0}{2}, \tag{4.11}$$

$$F_3 = y_2 + \left(3q y_0 - \frac{f_2}{2q}\right) y_1 + q^2 y_0^3 - \frac{f_2}{2} y_0^2 - \frac{f_1}{2} y_0 - \frac{f_0}{2}. \tag{4.12}$$

The restrictions ϕ to this dynamics are

$$\overline{\phi}_1 = A_0 y_1,$$

$$\overline{\phi}_2 = \frac{1}{2q} \left(-6q^3 y_0^3 + 6q^2 y_1 y_0 + 3q(f_2 y_0^2 + f_1 y_0 + f_0) - f_2 y_1\right),$$

$$\overline{\phi}_3 = \frac{1}{2q} \left(-6q^3 y_0^3 - 6q^2 y_1 y_0 + 3q(f_2 y_0^2 + f_1 y_0 + f_0) + f_2 y_1\right),$$

respectively.

As a result, we obtain the following theorem.

Theorem 4.2 *The FKPP equation*

$$u_t = u_{xx} + f_3 u^3 + f_2 u^2 + f_1 u + f_0, \tag{4.13}$$

with nonzero f_3, admits the following second-order dynamics of the form:

$$F = y_2 - A(y_0) y_1 - B(y_0).$$

– *If $f_3 > 0$, i.e. $f_3 = 2q^2$, then the dynamics has form (4.9);*
– *If $f_3 < 0$, i.e. $f_3 = -2q^2$, then the dynamics have forms (4.10)–(4.12).*

Here, q is a nonzero number.

4.4.2 Integration of the Dynamics

Consider, for example, dynamics (4.12). Corresponding differential equation has the form

$$y'' + \left(3qy - \frac{f_2}{2q}\right) y' + q^2 y^3 - \frac{f_2}{2} y^2 - \frac{f_1}{2} y - \frac{f_0}{2} = 0. \tag{4.14}$$

The distribution P is generated by the vector field

$$D = \frac{\partial}{\partial x} + y_1 \frac{\partial}{\partial y_0} - \left(\left(3q y_0 - \frac{f_2}{2q} \right) y_1 + q^2 y_0^3 - \frac{f_2}{2} y_0^2 - \frac{f_1}{2} y_0 - \frac{f_0}{2} \right) \frac{\partial}{\partial y_1}$$

(4.15)

or by the differential 1-forms

$$\omega_1 = dy_0 - y_1 dx,$$

$$\omega_2 = dy_1 + \left(\left(3q y_0 - \frac{f_2}{2q} \right) y_1 + q^2 y_0^3 - \frac{f_2}{2} y_0^2 - \frac{f_1}{2} y_0 - \frac{f_0}{2} \right) dx.$$

This distribution has two shuffling symmetries:

$$S_1 = \bar{\phi} \frac{\partial}{\partial y_0} + D(\bar{\phi}) \frac{\partial}{\partial y_1}$$

$$= \left(-3q^2 y_0^3 + \frac{3}{2} f_2 y_0^2 + \left(-3q y_1 + \frac{3}{2} f_1 \right) y_0 + \frac{1}{2q} f_2 y_1 + \frac{3}{2} f_0 \right) \frac{\partial}{\partial y_0}$$

$$\left(3q^3 y_0^4 - 2q f 2 y_0^3 + \frac{1}{4q^2} (-6q^3 f_1 + q f_2^2) y_0^2 + \frac{1}{4q^2} (q f_1 f_2 - 6q^3 f_0) y_0 \right.$$

$$\left. -3q y_1^2 + \frac{1}{4q^2} (6 f_1 q^2 + f_2^2) y_1 + \frac{1}{4q} f_0 f_2 \right) \frac{\partial}{\partial y_1}$$

and

$$S_2 = S_{y_1} = y_1 \frac{\partial}{\partial y_0} + D(y_1) \frac{\partial}{\partial y_1}$$

$$= y_1 \frac{\partial}{\partial y_0} + \left(-q^2 y_0^3 + \frac{1}{2} f_2 y_0^2 + \left(-3q y_1 + \frac{1}{2} f_1 \right) y_0 + \frac{1}{2q} f_2 y_1 + \frac{1}{2} f_0 \right) \frac{\partial}{\partial y_1}.$$

The vector fields S_1 and S_2 define commutative symmetry Lie algebra:

$$[S_1, S_2] = 0.$$

According to the Lie–Bianchi theorem [5, 11], the ordinary differential equation $F = 0$ is integrable by quadratures. In order to construct its first integrals, we construct two differential 1-forms ϖ_1 and ϖ_2 instead of the forms ω_1 and ω_2. We choose them so that they form a dual basis for the vector fields S_1 and S_2, i.e. $\varpi_i(S_j) = \delta_{ij}$, where δ_{ij} is the Kronecker delta. Compose the matrix

$$W = \begin{Vmatrix} \omega_1(S_1) & \omega_1(S_2) \\ \omega_2(S_1) & \omega_2(S_2) \end{Vmatrix}.$$

Determinant of the matrix is

$$\det W = \frac{1}{4q} \left(12q^5 y_0^6 + 36y_1 q^4 y_0^4 - 12y_0^2 (y_0^3 f_2 + f_1 y_0^2 + f_0 y_0 - 3y_1^2) q^3 \right.$$

$$- 18 \left(-\frac{2}{3} y_1^2 + f_1 y_0^2 + f_0 y_0 + \frac{4}{3} y_0^3 f_2 \right) q^2 y_1$$

$$+ \left(3f_2^2 y_0^4 + 6f_1 y_0^3 f_2 + (3f_1^2 + 6f_0 f_2) y_0^2 + (6f_0 f_1 - 12y_1^2 f_2) y_0 \right.$$

$$\left. + 3f_0^2 - 6y_1^2 f_1) q + 3y_1 f_2 (f_2 y_0^2 + f_1 y_0 + f_0) \right).$$

In the domain of the plane (y_0, y_1) where $\det W \neq 0$, there exists the inverse matrix W^{-1}. Define differential 1-forms ϖ_1 and ϖ_2:

$$\left\| \begin{matrix} \varpi_1 \\ \varpi_2 \end{matrix} \right\| = W^{-1} \left\| \begin{matrix} \omega_1 \\ \omega_2 \end{matrix} \right\|.$$

Since the Lie bracket $[S_1, S_2] = 0$, we get

$$d\varpi_i (S_1, S_2) = S_1(\varpi_i(S_2)) - S_2(\varpi_i(S_1)) - \varpi([S_1, S_2]) = 0.$$

This means that the forms ϖ_1 and ϖ_2 are closed. Due to the Poincaré lemma, there exist functions H_1 and H_2 such that $\varpi_1 = dH_1$ and $\varpi_2 = dH_2$. These functions are first integrals of the ordinary differential equation $F = 0$. Integrating the forms ϖ_1 and ϖ_2 along an arbitrary path in the space $J^1(\mathbb{R})$, we find these integrals. We do not write them for general case because of their bulkiness.

4.4.3 Construction Solutions of the FKPP Equation by Dynamics

To construct solutions of equation (4.13), we use Method 1 (see page 126).

Let $y(x; \mathbf{a})$, where $\mathbf{a} = (a_0, a_1)$, be the solution of ordinary differential equation (4.14) with initial conditions $y(x_0) = a_0$ and $y'(x_0) = a_1$. Let Φ_t be the shift transformation along the vector field S_1 from $t = 0$ to t. Since Φ_t is a symmetry of Eq. (4.14), the function $y(x; \Phi_t(\mathbf{a}))$ is a solution of this equation too.

The transformation Φ_t is defined by the solution of the ordinary equations

$$\begin{cases} \dfrac{dy_0}{dt} = -3q^2 y_0^3 + \dfrac{3}{2} f_2 y_0^2 + \left(-3qy_1 + \dfrac{3}{2} f_1\right) y_0 + \dfrac{1}{2q} f_2 y_1 + \dfrac{3}{2} f_0, \\[3mm] \dfrac{dy_1}{dt} = 3q^3 y_0^4 - 2qf 2 y_0^3 + \dfrac{1}{4q^2}(-6q^3 f_1 + qf_2^2) y_0^2 \\[3mm] \qquad + \dfrac{1}{4q^2}(qf_1 f_2 - 6q^3 f_0) y_0 - 3qy_1^2 + \dfrac{1}{4q^2}(6f_1 q^2 + f_2^2) y_1 + \dfrac{1}{4q} f_0 f_2 \end{cases}$$

$$(4.16)$$

with initial conditions $y_0(0) = y_0$ and $y_1(0) = y_1$.

Therefore, if we manage to solve this system and find the flow of the vector field S_1 in explicit form, then we can construct an exact solution of the FKPP equation. Otherwise, we can use numerical methods to system (4.16). As a result, we obtain approximate solutions of equation (4.13).

Example 4.1 Consider the equation

$$u_t = u_{xx} - 2u^3 + 1. \tag{4.17}$$

Then, $\phi = y_2 - 2y_0^3 + 1$, and we have three dynamics:

$$F_1 = y_2 + 3y_0 y_1 + y_0^3 - \frac{1}{2}; \tag{4.18}$$

$$F_2 = y_2 - 3y_0 y_1 + y_0^3 - \frac{1}{2}; \tag{4.19}$$

$$F_3 = y_2 - \alpha y_1 - 2y_0^3 + 1, \tag{4.20}$$

where α is a constant.

Consider dynamics (4.18), for example, i.e. suppose that $F = F_1$. The restriction of the function ϕ to the equation $F = 0$ is

$$\overline{\phi} = -3y_0^3 - 3y_0 y_1 + \frac{3}{2}.$$

The distribution $\mathbf{P}$ is generated by the differential 1-forms

$$\omega_1 = dy_0 - y_1 dx, \tag{4.21}$$

$$\omega_2 = dy_1 + \left(3y_0 y_1 + y_0^3 - \frac{1}{2}\right) dx. \tag{4.22}$$

The vector fields of shuffling symmetries are

$$S_1 = -\left(3y_0^3 + 3y_0 y_1 - \frac{3}{2}\right) \frac{\partial}{\partial y_0} + \left(-3y_1^2 + 3y_0^4 - \frac{3}{2} y_0\right) \frac{\partial}{\partial y_1}, \tag{4.23}$$

Fig. 4.1 The vector field S_1

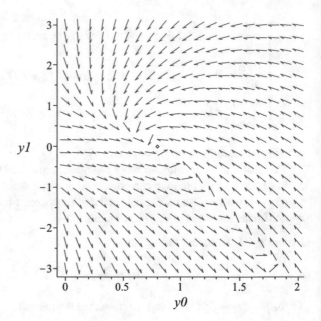

$$S_2 = y_1 \frac{\partial}{\partial y_0} - \left(3y_0 y_1 + y_0^3 - \frac{1}{2}\right) \frac{\partial}{\partial y_1}. \qquad (4.24)$$

Remark 4.1 The vector field S_1 has a stable focus at the point $y_0 = \frac{1}{\sqrt[3]{2}}$, $y_1 = 0$ (see Fig. 4.1).

The matrix W is

$$W = \begin{Vmatrix} -3y_0^3 - 3y_1 y_0 + \dfrac{3}{2} & y_1 \\ 3y_0^4 - 3y_1^2 - \dfrac{3}{2} y_0 & \dfrac{1}{2} - 3y_1 y_0 - y_0^3 \end{Vmatrix}.$$

It is nondegenerate if

$$\det W = 9y_0^2 y_1^2 - 3y_0^3 + 9y_0^4 y_1 + 3y_0^6 - \frac{9}{2}y_1 y_0 + 3y_1^3 + \frac{3}{4} \neq 0.$$

The differential 1-forms ϖ_1 and ϖ_2 are

$$\varpi_1 = -\frac{(6y_0 y_1 + 2y_0^3 - 1)dy_0 + 2y_1 dy_1}{2 \det W}, \qquad (4.25)$$

$$\varpi_2 = -dx - \frac{3\left((4y_1^2 - 4y_0^4 + 2y_0)dy_0 - 4\left(y_0 y_1 + y_0^3 - \frac{1}{2}\right)dy_1\right)}{4 \det W}. \qquad (4.26)$$

Fig. 4.2 Sections of the graph of $u(t, x)$ for $t = 0$ (red), 0.05 (blue), and 0.15 (green)

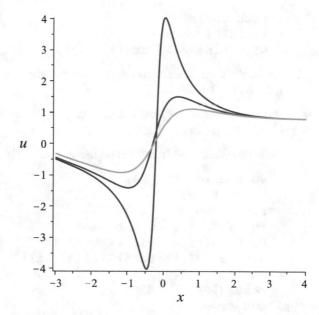

After integrating them, we get first integrals of the equation $F = 0$. However, these integrals are cumbersome and we do not give them here. Fortunately, a general solution of equation

$$y'' + 3yy' + y^3 - \frac{1}{2} = 0 \qquad (4.27)$$

can be constructed directly using Maple. Below we give the corresponding program code. As a result, we obtain the general solution of equation (4.27):

$$y(x) = \frac{C_1 e^{\frac{2}{\sqrt{3}}x} - \frac{1}{2}e^{-\frac{1}{\sqrt{3}}x}\left(\sqrt{3}C_2 + 1\right)\cos \chi + \frac{1}{2}e^{-\frac{1}{\sqrt{3}}x}\left(C_2 - \sqrt{3}\right)\sin \chi}{\sqrt[3]{2}\left(C_1 e^{\frac{2}{\sqrt{3}}x} - C_2 e^{-\frac{1}{\sqrt{3}}x}\sin \chi + e^{-\frac{1}{\sqrt{3}}x}\cos \chi\right)},$$

$$(4.28)$$

where C_1 and C_2 are arbitrary constants and $\chi = \frac{\sqrt{3}}{\sqrt[3]{16}}x$.

An example of calculations in Maple is given below.

Maple Code: Second-order dynamics for the equation $u_t = u_{xx} - 2u^3 + 1$

1. Load libraries:

```
with(DifferentialGeometry):
with(JetCalculus):
```

```
with(Tools):
with(PDETools):
with(LinearAlgebra):
```

2. Set jet notation, declare coordinates on the manifold M, and generate coordinates on the 3-jet space:

```
Preferences("JetNotation", "JetNotation2"):
DGsetup([x], [y], M, 3):
```

3. Define the Poisson–Lie bracket on the space $J^3(\mathbb{R})$:

```
Poisson:= proc (A, B)
local i, P;
P:=0:
for i from 0 to 3 by 1 do
P:=P+(diff(A, y[i])*TotalDiff(B, [i])-
diff(B, y[i])*TotalDiff(A, [i]))
end do:
return P:
end proc:
```

4. Define the function ϕ and a second-order dynamics F:

```
f(y[0]) := -2*y[0]^3+1:
phi := y[2]+f(y[0]):
F:=y[2]-A(y[0])*y[1]-B(y[0]):
```

5. The Poisson–Lie bracket calculation:

```
eq0:=collect(Poisson(phi,F),{y[1],y[2]}):
```

6. Substitution of the second derivative:

```
sub_y2:=y[2]=solve(F,y[2]):
```

7. Restriction of the Poisson–Lie bracket to the dynamics $F = 0$:

```
eq1:=[coeffs(collect(eval(eq0,sub_y2),y[1]),y[1])]:
```

8. Printing the resulting equations $\overline{[\phi, F]} = 0$:

```
for i from 1 to nops(eq1) by 1 do
print(simplify(eq1[i]))
end do;
```

9. Solve the resulting system $\overline{[\phi, F]} = 0$ with respect to the functions A and B:

```
dsolve(eq1);
```

10. We get dynamics (4.18)–(4.20). Choose dynamics (4.18):

```
F:=y[2]+3y[0]*y[1]+y[0]^3-1/2:
```

11. Convert the function F to a differential operator (equation)

    ```
    ode:=convert(F,DGdiff):
    ```

12. This equation can be solved by quadratures. Construct a solution of the Cauchy problem for this equation:

    ```
    Y:=simplify(unapply(rhs(dsolve({ode,y(0)=a1,
    (D(y))(0) = a2})),x,a1,a2)):
    ```

13. Restriction of ϕ to the dynamic $F = 0$:

    ```
    sub_y2:=y[2]=solve(F,y[2]):
    phi_F:=eval(phi, sub_y2):
    ```

14. Define vector field (4.15):

    ```
    Z:=evalDG(D_x+y[1]*D_y[0]+(rhs(sub_y2))*D_y[1]):
    ```

15. Define vector field (4.23):

    ```
    S1:=evalDG(phi_F*D_y[0]+LieDerivative(Z,phi_F)*D_y[1]):
    ```

16. To find the shifts Φ_t along the vector field S_1, we compose a system of differential equations

    ```
    z1:=diff(q(t),t)=eval(Hook(S1,dy[0]),
       {y[0]=q(t),y[1]=p(t)});
    z2:=diff(p(t),t)=eval(Hook(S1,dy[1]),
       {y[0]=q(t),y[1]=p(t)});
    ```

    Here, $q = y_0$ and $p = y_1$.

17. Choose the solution of equation (4.27) with initial data

    $$q(0) = a_0 = 4, \quad p(0) = a_1 = 2,$$

    and compose a system to calculate the shift of the point (a_0, a_1):

    ```
    ind:=q(0) = 4, p(0) = 2;
    dsys:={z1,z2,ind};
    ```

18. Numerically solve this system:

    ```
    dsn := dsolve(dsys, numeric);
    ```

19. Load the library:

    ```
    with(plots):
    ```

20. Form the image of sections of the solution of the equation at $t = 0; 0.05; 0.15$:

    ```
    r1:=plot(Y(x,rhs(dsn(0)[3]),rhs(dsn(0)[2])),
    x=-3..4,color="RED");
    r2:=plot(Y(x,rhs(dsn(0.05)[3]),rhs(dsn(0.1)[2])),
    ```

```
x=-3..4,color="Blue");
r3:=plot(Y(x,rhs(dsn(0.15)[3]),rhs(dsn(0.15)[2])),
x=-3..4,color="GREEN");
```

21. Display images on the screen:

```
display([r1,r2,r3],numpoints=1500,
resolution=3000,
thickness=2,axes = framed,
axesfont = ["TIMES", "ROMAN", 12],
labelfont = ["TIMES","ITALIC", 14],
labels = ["x", "y"],color="BLACK");
```

As a result, we obtain slices of the solution of the equation at moments $t = 0, 0.05, 0.15$ (see Fig. 4.2).

4.5   The Reaction–Diffusion Equation with a Convection Term

The reaction–diffusion equation with a nonlinear convection flow $H(u)$ in the positive direction of the x-axis has the form [14]

$$u_t + H(u)_x = u_{xx} + f(u). \tag{4.29}$$

Write this equation in the form

$$u_t = u_{xx} + g(u)u_x + f(u), \tag{4.30}$$

which is more convenient for calculations. Here, $g(u) = -H'(u)$. Then,

$$\phi(y_0, y_1, y_2) = y_2 + g(y_0)y_1 + f(y_0).$$

Below we suppose that

$$g' \neq 0. \tag{4.31}$$

4.5.1   First-Order Dynamics

Construct first-order dynamics of Eq. (4.30) in the following form:

$$F := y_1 - h(y_0) = 0, \tag{4.32}$$

where h is some smooth function.

The restriction of the Poisson–Lie bracket to Eq. (4.32) is

$$\overline{[\phi, F]} = h'f - hf' - h^2(g' + h'').$$

Equation $\overline{[\phi, F]} = 0$ has the trivial solution $h = 0$, which corresponds to x-independent solutions of equation (4.30). Consider the case when $h \neq 0$. Then, the function h satisfies the Abel differential equation of second kind (see [1])

$$hh' + (g(y_0) + \alpha)h + f(y_0) = 0, \tag{4.33}$$

where α is a constant. Due to (4.33), the evolutionary vector field has the form

$$S = \left(hh' + gh + f\right) \frac{\partial}{\partial y_0} = \alpha h \frac{\partial}{\partial y_0}.$$

4.5.2 Second-Order Dynamics

We will look for second-order dynamics in the form of the Liénard equation too (see (4.8)). The Poisson–Lie bracket is

$$[\phi, F] = -(g''+A'')y_1^3 -(f''+B'')y_1^2 -\left(2(g' + A')y_2 - g'B + A'f\right)y_1 + B'f - f'B.$$

Since its restriction to Eq. (4.8) is a polynomial in y_1, Eq. (4.6) implies the following system of ordinary differential equation:

$$\begin{cases} (f - 2B)A' - 3g'B = 0, \\ 2(A' + g')A + f'' + B'' = 0, \\ B'f - Bf' = 0, \\ g'' + A'' = 0. \end{cases}$$

Solving this system, we find that the functions g and f should be linear and cubic, respectively;

$$f(y_0) = f_3 y_0^3 + f_2 y_0^2 + f_1 y_0 + f_0, \tag{4.34}$$

$$g(y_0) = g_1 y_0 + g_0, \tag{4.35}$$

where $f_0, \ldots, f_3, g_0, g_1 \in \mathbb{R}$. From inequality (4.31), it follows that $g_1 \neq 0$.

Theorem 4.3 *1. Equation (4.30) has second order finite dimensional dynamics in the form of the Liénard equation (4.8) if and only if the function f is a polynomial of third degree and the function g is linear.*

2. Suppose that the functions f and g have forms (4.34) and (4.35), respectively, where $f_3 \neq 0$, $g_1 \neq 0$. Then Eq. (4.30) has the finite dimensional dynamics

$$F = y_2 + (g_1 y_0 + \alpha) y_1 + f_3 y_0^3 + f_2 y_0^2 + f_1 y_0 + f_0,$$

where α is a constant. In addition, if the condition $g_1^2 - 8 f_3 \geq 0$ holds, then Eq. (4.30) has one more finite dimensional dynamics

$$F = y_2 - (A_1 y_0 + A_0) y_1 + \frac{1}{3} (A_1^2 + g_1 A_1 + 3 f_3) y_0^3$$
$$+ (A_1 A_0 + f_2 + g_1 A_0) y_0^2 - B_1 y_0 - B_0,$$

where

$$A_0 = \frac{f_2 \beta}{f_3}, \quad A_1 = 3\beta, \quad B_0 = \frac{f_0(f_3 + g_1 \beta)}{2 f_3},$$

$$B_1 = \frac{f_1(f_3 + g_1 \beta)}{2 f_3}, \quad \beta = \frac{-g_1 \pm \sqrt{g_1^2 - 8 f_3}}{4}.$$

Example 4.2 Consider the equation

$$u_t = u_{xx} - (u + 1) u_x + \frac{1}{8} u^3. \tag{4.36}$$

Then,

$$\phi = y_2 - (y_0 + 1) y_1 + \frac{1}{8} y_0^3$$

and

$$F = y_2 - \frac{3}{4} y_0 y_1 + \frac{1}{16} y_0^3.$$

Restrict ϕ to the equation $F = 0$:

$$\overline{\phi} = -\frac{1}{4} y_0 y_1 + \frac{1}{16} y_0^3 - y_1.$$

The vector field is

Fig. 4.3 The graph of
solution (4.38)

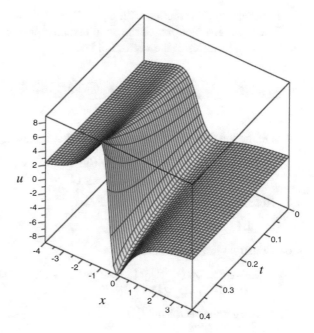

$$S = \left(-\frac{1}{4}y_0 y_1 + \frac{1}{16}y_0^3 - y_1\right)\frac{\partial}{\partial y_0} + \left(-\frac{1}{4}y_1^2 - \frac{3}{4}y_0 y_1 + \frac{1}{64}y_0^4 + \frac{1}{16}y_0^3\right)\frac{\partial}{\partial y_1}.$$

The shift transformation Φ_t corresponding to this field is

$$x \longmapsto x,$$

$$y_0 \longmapsto \frac{8(ty_0^2 - 4y_1 t + 4y_0)}{(t^2 - 2t)y_0^2 + 8ty_0 + 32 + (8t - 4t^2)y_1},$$

$$y_1 \longmapsto \frac{8((2t+t^2)y_0^4 + 8ty_0^3 - 8ty_1(2+t)y_0^2 - 32ty_0 y_1 + 128y_1 + 16t^2 y_1^2 + 32ty_1^2)}{\left((t^2 - 2t)y_0^2 + 8ty_0 - 4t^2 y_1 + 8y_1 t + 32\right)^2}.$$

The equation $F = 0$ has the following general solution:

$$y(x) = -\frac{8(C_1 x + C_2)}{C_1 x^2 + 2C_2 x + 2}, \tag{4.37}$$

where C_1 and C_2 are arbitrary constants. Applying the inverse transformation Φ_t^{-1}
to these functions, we get a solution of equation (4.36):

$$u(t, x) = -\frac{8(C_1(x - t) + C_2)}{2 + ((x - t)^2 - 2t)C_1 + (2(x - t))C_2}. \tag{4.38}$$

The graph of this solution with $C_1 = C_2 = 1$ is shown in Fig. 4.3.

Below we give the part of the code responsible for the shift of solutions (4.37) along the trajectories of the vector field S.

Maple Code: Second-order dynamics for Eq. (4.36)

4. Define the function ϕ and a second-order dynamics F:

```
phi:=y[2]-(y[0]+1)*y[1]+(1/8)*y[0]^3:
F  :=  y[2]-(3/4)*y[0]*y[1]+(1/16)*y[0]^3:
```

5. Restriction of ϕ to the dynamic $F = 0$:

```
phi_F:= phi-F:
```

6. Construct the vector field $\mathcal{D}$:

```
Z:=evalDG(D_x+y[1]*D_y[0]+(A(y[0])*y[1]+
B(y[0]))*D_y[1]):
```

7. Construct the vector field $\overline{S}$:

```
S:=evalDG(phi_F*D_y[0]+
LieDerivative(Z,phi_F)*D_y[1]):
```

8. The transformation Φ_t and its inverse transformation Φ_t^{-1}:

```
Phi:=Flow(S,t):
Xi:=InverseTransformation(Phi):
```

9. Solution (4.37) and its derivative:

```
ode:=convert(F1,DGdiff): v:=rhs(dsolve(ode));
w:=diff(v,x);
```

10. Apply the transformation Φ_t^{-1} to solution (4.37):

```
eq:=Pullback(Xi,[y[0]-v,y[1]-w]);
```

11. Find the explicit form of the solution:

```
sol:=solve(eq,{y[0],y[1]}):
q:=rhs(sol[1]);
```

12. Checking:

```
U:=diff(u(t,x),t)-eval(convert(phi,DGdiff),
y(x)=u(t,x));
simplify(eval(U,u(t,x)=q));
```

4.6 The Black–Scholes Equation

The Black–Scholes equation

$$u_t = -\frac{1}{2}\sigma^2 x^2 u_{xx} - rx u_x + ru \tag{4.39}$$

is a well-known linear partial differential equation of financial mathematics [4]. It describes the price of the option over time. Here, u is the price of the option as a function of stock price x and time t, r is the risk-free interest rate, and σ is the volatility of the stock. Note that, unlike the equations considered above, this equation depends on the variable x. For this equation,

$$\phi = -\frac{1}{2}\sigma^2 x^2 y_2 - rx y_1 + ry_0.$$

4.6.1 First-Order Dynamics

Since the equation is linear, we will seek its linear dynamics:

$$F = y_1 - A(x)y_0 - B(x). \tag{4.40}$$

Then,

$$[\phi, F] = \left(\frac{\sigma^2 x^2}{2}A''y_0 + rxA'\right)y_0 + \left(r + \sigma^2 x^2 A'\right)y_1$$

$$+ \sigma^2 xy_2 + \frac{\sigma^2 x^2}{2}B'' + rxB' - rB.$$

The restriction $\overline{[\phi, F]}$ of this bracket in Eq. (4.40) is a linear function with respect to y_0. Therefore, the equation $\overline{[\phi, F]} = 0$ is equivalent to the following system of two ordinary differential equations:

$$\begin{cases} \dfrac{1}{2}A''\sigma^2 x^2 + (A\sigma^2 x + \sigma^2 + r)xA' + A^2\sigma^2 x + Ar = 0, \\[2mm] \dfrac{1}{2}B''\sigma^2 x + (\sigma^2 + r)B' + \sigma^2(xA' + A)B = 0. \end{cases}$$

The first equation can be solved:

$$A(x) = \frac{1}{2\sigma^2 x}\left(\sigma^2 - 2r - C_1 \tan\left(\frac{C_1 \ln x - C_2}{2\sigma^2}\right)\right),$$

where C_1 and C_2 are arbitrary constants. After substituting into the second equation, we obtain the equation for B:

$$4x\left(\frac{\sigma^2 x}{2}B'' + (\sigma^2 + r)B'\right)\sigma^2 \cos^2\left(\frac{C_1 \ln x - C_2}{2\sigma^2}\right) - BC_1^2 = 0. \tag{4.41}$$

It is quite difficult to construct a general solution of this equation. But if we put $C_1 = 0$, then it can be easily solved:

$$B(x) = C_3 x^{-\frac{\sigma^2 + 2r}{\sigma^2}} + C_4,$$

where C_3 and C_4 are arbitrary constants. Then, we get the following dynamics:

$$F = y_1 - \frac{1}{2\sigma^2 x}(\sigma^2 - 2r)y_0 - C_3 x^{-\frac{\sigma^2 + 2r}{\sigma^2}} - C_4. \tag{4.42}$$

The general solution of the corresponding equation is

$$y(x) = C_5 x^{\frac{\sigma^2 - 2r}{2\sigma^2}} - \frac{2\sigma^2(C_3 x^{-\frac{2r}{\sigma^2}} - C_4 x)}{\sigma^2 + 2r}, \tag{4.43}$$

where C_5 is an arbitrary constant. A zero solution $B(x) = 0$ of Eq. (4.41) gives another dynamics:

$$F = y_1 - \frac{1}{2\sigma^2 x}\left(\sigma^2 - 2r - C_1 \tan\left(\frac{C_1 \ln x - C_2}{2\sigma^2}\right)\right)y_0. \tag{4.44}$$

Its general solution is

$$y(x) = C_3 x^{\frac{\sigma^2 - 2r}{2\sigma^2}} \cos\left(\frac{C_1 \ln x - C_2}{2\sigma^2}\right). \tag{4.45}$$

4.6.2 Construction Solutions of the Black–Scholes Equation by Dynamics

At first, consider dynamics (4.44). Restrict the function ϕ to dynamics (4.44):

$$\overline{\phi} = \frac{C_1^2 + (\sigma^2 + 2r)^2}{8\sigma^2}y_0.$$

Construct the vector field

$$\overline{S} = \overline{\phi} \frac{\partial}{\partial y_0},$$

and find its shift transformation:

$$\Phi_t : (x, y_0) \longmapsto \left(x, e^{\frac{C_1^2 + (\sigma^2 + 2r)^2}{8\sigma^2} t} y_0 \right).$$

The inverse transformation is

$$\Phi_t^{-1} : (x, y_0) \longmapsto \left(x, e^{-\frac{C_1^2 + (\sigma^2 + 2r)^2}{8\sigma^2} t} y_0 \right).$$

Acting by this transformation on function (4.45), we obtain the following exact solution of equation (4.39):

$$u(t, x) = C_3 x^{\frac{\sigma^2 - 2r}{2\sigma^2}} \cos\left(\frac{C_1 \ln x - C_2}{2\sigma^2} \right) e^{-\frac{C_1^2 + (\sigma^2 + 2r)^2}{8\sigma^2} t}. \tag{4.46}$$

Here, C_1, C_2, and C_3 are arbitrary constants.

For example, the function

$$u(t, x) = \frac{e^{\frac{3}{2} t}}{\sqrt{2x}} \sqrt{\sin(\sqrt{3} \ln x) + 1} \tag{4.47}$$

is a solution of equation (4.39) with $\sigma = r = 1$ (see Figs. 4.4 and 4.5).

Now, consider dynamics (4.42). In this case,

$$\overline{\phi} = \frac{1}{8\sigma^2} \left(2r + \sigma^2 \right) \left((2r + \sigma^2) y_0 + 2\sigma^2 C_3 x^{-\frac{2r}{\sigma^2}} - 2C_4 \sigma^2 x \right).$$

Omitting cumbersome calculations, we write the final result. The function

$$u(t, x) = B_1 x + B_2 x^{-\frac{2r}{\sigma^2}} + B_3 x^{\frac{\sigma^2 - 2r}{2\sigma^2}} e^{\frac{(\sigma^2 + 2r)^2}{8\sigma^2} t}$$

is a solution of equation (4.39). Here, B_1, B_2, and B_3 are arbitrary constants.

Fig. 4.4 Sections of the
graph of solution (4.47) for
$t = 0$ (red), 0.1 (orange), 0.2
(green), and 0.3 (blue)

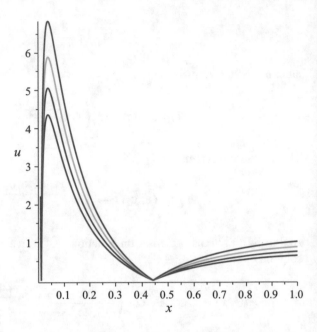

Fig. 4.5 The graph of
solution (4.47)

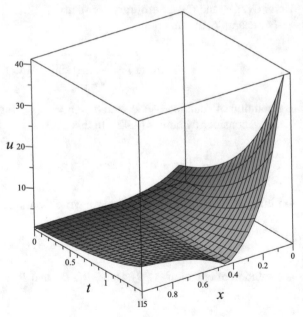

Maple Code: First-order dynamics for the Black–Scholes equation

The first three items are the same as in the Maple code on page 135. We start at the fourth one.

4. Define the function ϕ and a second-order dynamics F:

```
phi := -sigma^2*x^2*y[2]/2-r*x*y[1]+r*y[0]:
F := y[1]-A(x)*y[0]-B(x):
```

5. The Poisson–Lie bracket calculation:

```
eq0:=simplify(Poisson(phi,F),size);
```

6. Substitution of the derivatives:

```
sub:=solve([F,TotalDiff(F,x)],{y[1],y[2]}):
```

7. Restriction of the Poisson–Lie bracket to the dynamics:

```
eq1:=simplify(eval(eq0,sub),size):
eq2:=[coeffs(collect(eq1,y[0],distributed),y[0])];
```

8. Print the resulting equations $\overline{[\phi, F]} = 0$:

```
for i from 1 to nops(eq2) by 1 do
print(simplify(eq2[i],size))
end do;
```

9. Next, the resulting system of equations is solved with respect to the function A and B in a semi-manual mode. As a result, we get dynamics (4.44):

```
F:=y[1]-(1/2)*(sigma^2-2*r
-tan((1/2)*C1*(ln(x)-C2)/sigma^2)*C1)*y[0]/
(sigma^2*x):
```

10. Restriction of ϕ to the dynamic $F = 0$:

```
phi_F:=simplify(eval(phi,sub),size):
```

11. Construct the vector field $\overline{S}$:

```
S:=evalDG(phi_F*D_y[0]):
```

12. The transformation Φ_t and its inverse transformation Φ_t^{-1}:

```
Phi:=Flow(S,t):
Xi:=InverseTransformation(Phi):
```

13. Apply the transformation Φ_t^{-1} to solution (4.45):

```
Xi_y:=Pullback(Xi,y[0]-(C3*x^((1/2)
*(sigma^2-2*r)/sigma^2)*cos((C1*ln(x)-C2)
/(2*sigma^2)))):
```

14. Find solution (4.46):

    ```
    q:=solve(Xi_y,y[0]);
    ```

15. Check this solution:

    ```
    BSch:=diff(u(t,x),t)-eval(convert(phi,DGdiff),
    y(x)=u(t,x)):
    simplify(eval(BSch,u(t,x)=q));
    ```

    It should be zero.

Acknowledgments This work is partially supported by Russian Foundation for Basic Research (project 18-29-10013).

References

1. N. H. Abel, Précis d'une théorie des fonctions elliptiques, *J. Reine Angew. Math.*, 4(1829), 309.348.
2. Akhmetzyanov A. V., Kushner A. G., Lychagin V. V., Attractors in Models of Porous Media Flow, *Doklady. Mathematics* **472**:6 (2017), 627–630.
3. Salnikov A. M., Akhmetzianov A.V., Kushner A.G., Lychagin V.V., A Numerical Method for Constructing Attractors of Evolutionary Filtration Equations, *2019 1st International Conference on Control Systems, Mathematical Modelling, Automation and Energy Efficiency (SUMMA)*, Lipetsk, Russia (2019), pp. 22–24, https://doi.org/10.1109/SUMMA48161.2019. 8947585.
4. Black F., Scholes M., The Pricing of Options and Corporate Liabilities, *The Journal of Political Economy* **81**:3 (1973), 637–654.
5. Duzhin S. V., Lychagin V. V., Symmetries of distributions and quadrature of ordinary differential equations, *Acta Appl. Math.* **24** (1991), 29–57.
6. Fisher R. A. The wave of advance of advantageous genes, *Annals of Eugenics.* **7** (1937), 353–369.
7. Kruglikov B. S., Lychagina O. V., Finite dimensional dynamics for Kolmogorov – Petrovsky – Piskunov equation, *Lobachevskii Journal of Mathematics* **19** (2005), 13–28.
8. A.N. Kolmogorov, I.G. Petrovskii, N.S. Piskunov, A study of diffusion with increase in the quantity of matter, and its application to a biological problem, *Bull. Moscow State Univ.* **17** (1937), 1–72.
9. Krasilshchik I. S., Lychagin V. V., Vinogradov A. M., Geometry of jet spaces and nonlinear partial differential equations, New York: Gordon and Breach, 1986.
10. Kushner A. G., Matviichuk R.I., Exact solutions of the Burgers – Huxley equation via dynamics, *Journal of Geometry and Physics* **151** (2020)
11. Kushner A. G., Lychagin V. V., Rubtsov V. N., Contact geometry and nonlinear differential equations, *Encyclopedia of Mathematics and Its Applications,* **101**. Cambridge: Cambridge University Press, xxii+496 pp., 2007.
12. Liénard A., Etude des oscillations entretenues, *Revue générale de l'électricité* **23** (1928), 901–912 and 946–954.
13. Lychagin V. V., Lychagina O. V., Finite Dimensional Dynamics for Evolutionary Equations, *Nonlinear Dyn.*, **48** (2007), 29–48.
14. Murray J. D., Mathematical Biology, Springer-Verlag, New York 1993.

15. NagumoJ., Yoshizawa S., and Arimoto S., Bistable transmission lines, *Circuit Theory, IEEE Transactions on*, **12**,no 3 (1965), 400–412.
16. Nourazar S.S., Soori M. and Nazari-Golshan A., On the Exact Solution of Burgers – Huxley Equation Using the Homotopy Perturbation Method, *Journal of Applied Mathematics and Physics* 3 (2015), 285–294.
17. Osipov V.V., The simplest autowaves, *Soros Educational Journal*, **7** (1999), 115–121.
18. Satsuma J. Exact solutions of Burgers equation with reaction terms. In: Topics in soliton theory and exactly solvable nonlinear equations. (M.J. Ablowitz, B. Fuchssteiner, M.D. Kruskal eds). Singapore: World Scientific, 1987. pp. 255–262.
19. Singh B., Kumar R., Exact Solutions of Certain Nonlinear Diffusion-Reaction Equations with a Nonlinear Convective Term, *International Journal of Pure and Applied Physics* **13**:1 (2017), 125–132.
20. Tychkov S.N. Introduction to Symbolic Computations in Differential Geometry with Maple, *Nonlinear PDEs, Their Geometry, and Applications*, Birkhäuser, Cham (2019). 157–182.
21. Zeldovich Y., Frank-Kamenetskii D.A., A theory of thermal propagation of flame, *Acta Physicochimica USSR*, **9** (1938) 341–350.

Chapter 5
Critical Phenomena in Darcy and Euler Flows of Real Gases

Valentin V. Lychagin and Mikhail D. Roop

5.1 Introduction

In this paper, we summarize the results obtained in [1–5] for two important types of flows of real gases—the flows through porous media [6] and Euler flows.

The first significant results in the field of filtration processes were obtained in [7] and [8]. In these papers, the Darcy law was suggested to investigate flows through porous media. This approach appeared to be sufficiently effective. Such phenomena as phase transitions in filtration processes were studied in a few works, for instance, in [9] and [10]. In [9], one-dimensional nonstationary filtration of two-component mixture of hydrocarbons described by the generalized van der Waals equation was studied, but only one (thermic) equation of state was considered. In [10], the authors investigated non-equilibrium phase transitions in filtration of gas-condensate mixtures and provided the comparison with equilibrium phase transitions. In both works [9] and [10] numerical computations were used.

Comparing with [9] and [10], we consider three-dimensional, stationary, one-component filtration and provide explicit formulae for finding solutions of the Dirichlet boundary problem. Some exact solutions for nonstationary filtration together with the analysis of the symmetry algebra of corresponding equations for various media are presented in [11].

Navier–Stokes and Euler flows formed by means of source have been of great interest since the middle of the 20th century, when Landau found a new exact singular solution of incompressible Navier–Stokes equations [12] called *submerged jet*. From the physical viewpoint, this solution is formed by a source that transmits the momentum to surrounding medium in a certain direction. It has two features

V. V. Lychagin · M. D. Roop (✉)
Institute of Control Sciences of Russian Academy of Sciences, Moscow, Russia
e-mail: valentin.lychagin@uit.no; mihail_roop@mail.ru

© The Author(s), under exclusive license to Springer Nature Switzerland AG 2021
M. Ulan, E. Schneider (eds.), *Differential Geometry, Differential Equations, and Mathematical Physics*, Tutorials, Schools, and Workshops in the Mathematical Sciences, https://doi.org/10.1007/978-3-030-63253-3_5

that are usually said to be the drawbacks of Landau's solution. The first one is the triviality of solution in case of ideal (inviscid) fluids, which means that solution of this type is valid only for highly viscous fluids. The second one is zero mass flux through a sphere around the source. Landau's solution was improved by Broman and Rudenko [13]. By means of symmetry methods [14, 15], they constructed some exact solutions of incompressible Navier–Stokes system corresponding to flows with non-zero mass flux and valid for inviscid fluids. It is worth to say that in [12] and [13], thermodynamics was out of consideration. One of the first authors who investigated thermodynamical properties of such flows was Squire [16]. Taking into account the equation of heat balance, he found the distribution of the temperature in the jet for incompressible fluids. Some singular solutions for incompressible and compressible fluids invariant with respect to subalgebras of the symmetry algebra of the Navier–Stokes equations were obtained in [17]. In [18], one-dimensional gas flow with phase transitions was studied, and the van der Waals equation was chosen as a model of thermodynamic state.

For modelling flows formed by means of source, we use Euler equations extended by equations of state of the medium. In case of one point source, we provide a method of finding corresponding solutions for various gases and thermodynamical processes.

This paper has the following structure. In Sect. 5.2, we show that thermodynamics is a particular case of measurement of random vectors [19]; namely, thermodynamics can be considered as a theory of measurement of extensive thermodynamical variables. Such a consideration leads us to the geometric formulation of thermodynamics previously established, for instance, in [20] an [21], but we decided to include this part, on the one hand, to show a new approach to thermodynamics based on measurement, and on the other hand, to make the paper self-contained. In Sect. 5.3, we discuss thermodynamics of gases and show how methods of contact and symplectic geometry can help us, in particular, in solution of practical issues concerning the determination of caloric equation of state for gases if we know a thermic one and description of phase transitions. In Sect. 5.4, we illustrate the methods developed on concrete models of real gases, paying special attention to phase transitions. In Sect. 5.5, we discuss basic equations describing filtration processes and Euler flows. We provide two theorems that give us solutions for the corresponding problems that can be applied to any gas as well as any process. In Sect. 5.6, we provide solutions for gases discussed in Sect. 5.4 and analyze phase transitions along the corresponding flows. In Sect. 5.7, we discuss the results.

5.2   Measurement and Thermodynamics

In this section, we describe the measurement of random vector procedure and show that the results can be naturally presented in terms of contact and symplectic geometries. The more comprehensive discussion can be found in [19].

5.2.1   Measurement of Random Vectors

Let $(\Omega, \mathcal{A}, p)$ be a probability space, i.e. Ω is a set, $\mathcal{A}$ is a σ-algebra and p is a probability measure. By a random vector, we shall mean the following map: $X: (\Omega, \mathcal{A}, p) \to W$, where W is a vector space, $\dim(W) = n$.

Let us suppose that we have a device that measures random vector X. It is natural to require that the result produced by such a device will be an *expectation*

$$\mathbb{E}(X) = \int\limits_{\Omega} X dp,$$

where the integration is assumed to be coordinate-wise. This is what we mean by a *measurement*.

Due to Jensen's inequality, it is easy to check that in this case, the expectation of length of vector $X - c$ for some $c \in W$ with respect to any metric g on W reaches its minimal value. Indeed,

$$\mathbb{E}\left(g(X - c, X - c)\right) \geq g(\mathbb{E}(X - c), \mathbb{E}(X - c)) = g(\mathbb{E}(X) - c, \mathbb{E}(X) - c) \geq 0,$$

and the equality to zero takes place iff $c = \mathbb{E}(X)$.

5.2.2   Information Gain

Let us now suppose that we have another probability measure q on our probability space $(\Omega, \mathcal{A}, p)$, which has the same set of measure zero sets as p has. Such measures are called *equivalent*, and we will denote it by $p \sim q$.

Define the so-called «surprise function» $s: \mathcal{A} \ni A \longmapsto s(A) \in \mathbb{R}$ by the following formula:

$$s(A) = -\ln(p(A)).$$

Note that this function satisfies the following properties:

1. $s(\Omega) = 0$, $s(\varnothing) = \infty$ and $s(A) \geq 0$;
2. if A and B are independent, i.e. $p(AB) = p(A)p(B)$, then $s(AB) = s(A) + s(B)$;
3. s is a continuous function of $p(A)$.

Properties (1)–(3) may also serve as a definition of the surprise function.

Let $\Omega = \{\omega_1, \ldots, \omega_n\}$ be a finite set and $p = \{p_1, \ldots, p_n\}$, $p_i = p(\omega_i)$ be a probability measure. In this case, the expectation of the surprise function is

$$S(p) = -\sum_{i=1}^{n} p_i \ln p_i.$$

The above formula coincides with Shannon's definition of entropy. Let $q = \{q_1, \ldots, q_n\}$ now be another probability measure equivalent to p. Then, the changing of the surprise function will be

$$s(p, q) = s(q) - s(p) = \left(-\sum_{i=1}^{n} \ln q_i\right) - \left(-\sum_{i=1}^{n} \ln p_i\right) = \sum_{i=1}^{n} \ln\left(\frac{p_i}{q_i}\right).$$

And the average $I(p, q)$ of $s(p, q)$ with respect to the measure p that is called *gain of information* will be

$$I(p, q) = \sum_{i=1}^{n} p_i \ln\left(\frac{p_i}{q_i}\right).$$

If $\Omega = \mathbb{R} = \bigcup_i [x_i, x_{i+1}]$ and $dp = f(x)dx$, $dq = g(x)dx$, then $I(p, q)$ takes the following form:

$$I(p, q) \approx \sum_i f(\xi_i) \Delta_i \ln\left(\frac{f(\xi_i)}{g(\xi_i)}\right),$$

where $\Delta_i = x_{i+1} - x_i$. Taking limit $\Delta_i \to 0$, $i \to \infty$, one gets

$$I(p, q) = \int_{\mathbb{R}} f(x) \ln\left(\frac{f(x)}{g(x)}\right) dx.$$

In case of arbitrary probability space $(\Omega, \mathcal{A}, p)$, the gain of information $I(p, q)$ is defined by the formula

$$I(p, q) = \int_{\Omega} \ln\left(\frac{dp}{dq}\right) dp. \tag{5.1}$$

The function $I(p, q)$ has the property $I(p, q) \geq 0$ and $I(p, q) = 0$ iff $p = q$ almost everywhere. At the same time, it cannot serve as a distance between measures p and q since it is not symmetric, i.e. $I(p, q) \neq I(q, p)$ and does not satisfy the triangle inequality.

In terms of density ρ, such that $dp = \rho dq$, (5.1) can be written as

$$I(\rho) = \int_{\Omega} \rho \ln \rho \, dq.$$

5.2.3 The Principle of Minimal Information Gain

Let $x \in W$ be a fixed vector which is expected to be the result of the measurement of random vector $X : (\Omega, \mathcal{A}, q) \to W$, i.e.

$$\mathbb{E}(X) = x.$$

Obviously, the given measure q may not give us the required vector x. This means that we should choose another measure p, such that $dp = \rho dq$. In other words, we are looking for a function ρ such that

$$\int_{\Omega} \rho dq = 1, \quad \int_{\Omega} \rho X dq = x. \tag{5.2}$$

Conditions in (5.2) are not enough to determine ρ. In addition to (5.2), we require that the new measure p is the closest one to the measure q with respect to the gain of information $I(\rho)$, i.e.

$$I(\rho) = \int_{\Omega} \rho \ln \rho dq \to \min_{\rho}. \tag{5.3}$$

This is exactly what is called *the principle of minimal information gain*.

Thus, we have the following extremal problem. One needs to find the function ρ minimizing functional (5.3) under constraints in (5.2).

Theorem 5.1 *The solution of (5.2)–(5.3) is given by the following formulae:*

$$\rho = \frac{1}{Z(\lambda)} e^{\langle \lambda, X \rangle}, \quad Z(\lambda) = \int_{\Omega} e^{\langle \lambda, X \rangle} dq, \tag{5.4}$$

where $\lambda \in W^*$. The results of the measurement with respect to extremal measure p belong to a manifold

$$L_H = \left\{ x = -\frac{\partial H}{\partial \lambda} \right\} \subset W \times W^*,$$

where $H(\lambda) = -\ln Z(\lambda)$.

Remark 5.1

1. The measure p defined by relations in (5.4) is called *the extremal measure*.
2. The function $Z(\lambda)$ is called *the partition function* and, obviously, exists iff $\langle \lambda, X \rangle \leq 0$ almost everywhere.

3. The integral $Z(\lambda) = \int\limits_{\Omega} e^{\langle \lambda, X \rangle} dq$ can be expressed in terms of vector space W only:

$$Z(\lambda) = \int\limits_{W} e^{\langle \lambda, t \rangle} d\mu(t),$$

where $\mu = X_*(q)$.

Proof Consider the functional

$$\mathcal{L} = \int\limits_{\Omega} \rho \ln \rho \, dq - \lambda_0 \left(\int\limits_{\Omega} \rho \, dq - 1 \right) - \left\langle \lambda, \int\limits_{\Omega} \rho X dq - x \right\rangle.$$

Since its first variation with respect to ρ should be equal to zero, we get

$$\delta \mathcal{L} = \int\limits_{\Omega} (\ln \rho + 1 - \lambda_0 - \langle \lambda, X \rangle) \delta \rho \, dq = 0,$$

from what follows that

$$\rho = \exp(\lambda_0 - 1 + \langle \lambda, X \rangle).$$

Taking into account that $\int\limits_{\Omega} \rho \, dq = 1$, we get

$$\rho = \frac{1}{Z(\lambda)} e^{\langle \lambda, X \rangle}, \text{ where } Z(\lambda) = \int\limits_{\Omega} e^{\langle \lambda, X \rangle} dq.$$

Note that

$$\frac{\partial Z}{\partial \lambda} = \int\limits_{\Omega} X e^{\langle \lambda, X \rangle} dq = \int\limits_{\Omega} X \rho Z(\lambda) dq = Z(\lambda) x,$$

from what follows that

$$\frac{\partial}{\partial \lambda} (\ln Z(\lambda)) = x.$$

Introducing the Hamiltonian $H(\lambda) = -\ln Z(\lambda)$, we get

$$x = -\frac{\partial H}{\partial \lambda}. \tag{5.5}$$

One can see that the manifold $L_H \subset (\Phi, \omega)$ is Lagrangian with respect to nondegenerate closed 2-form on $\Phi = W \times W^*$

$$\omega = d\lambda \wedge dx = \sum_{i=1}^{n} d\lambda_i \wedge dx_i,$$

i.e. $\omega|_{L_H} = 0$.

A pair (Φ, ω) represents the standard model of symplectic space. Moreover, the Lagrangian manifold L_H gives us information about both extreme measure p and expectation of random vector X; namely, λ is responsible for the corresponding extremal measure, while x represents the expectation.

Let us introduce a new function

$$J(\lambda, x) = H(\lambda) + \langle \lambda, x \rangle.$$

Using Theorem 5.1, it is easy to show that there is a following relation between I and J:

$$J|_{L_H} = I. \tag{5.6}$$

Let us consider the differential of J:

$$dJ = \sum_i \left(\frac{\partial H}{\partial \lambda_i} + x_i \right) d\lambda_i + \sum_i \lambda_i dx_i = \sum_i \left(\frac{\partial H}{\partial \lambda_i} + x_i \right) d\lambda_i + \theta,$$

where 1-form θ has the following structure:

$$\theta = \sum_i \lambda_i dx_i.$$

On the surface L_H, we have

$$dJ|_{L_H} = \theta|_{L_H}. \tag{5.7}$$

From (5.6) and (5.7), it follows that

$$\theta|_{L_H} = dI. \tag{5.8}$$

Now, we construct the contactization $\tilde{\Phi}$ of Φ as

$$\tilde{\Phi} = \mathbb{R} \times \Phi = \mathbb{R} \times W \times W^* = \mathbb{R}^{2n+1}(u, x, \lambda)$$

and equip $\tilde{\Phi}$ with the contact 1-form

$$\tilde{\theta} = du - \theta. \tag{5.9}$$

Thus, $(\tilde{\Phi}, \tilde{\theta})$ is a contact space.

Let $a = (\lambda, x) \in L_H$, and let $\tilde{L} \subset \tilde{\Phi}$ be a submanifold of dimension n such that

$$\tilde{L} = \left\{ u = I(a), \quad x = -\frac{\partial H}{\partial \lambda} \right\}. \tag{5.10}$$

Note that $\tilde{L}$, being constructed in such a way, becomes a Legendrian submanifold, i.e.

$$\tilde{\theta}|_{\tilde{L}} = 0.$$

Indeed, due to (5.8),

$$\tilde{\theta}|_{\tilde{L}} = (du - \theta)|_{\tilde{L}} = dI - \theta|_{L_H} = 0.$$

Moreover, the Legendrian manifold $\tilde{L}$ provides the knowledge about not only extreme measure p and expectation of random vector X but also information gain $I(p, q)$.

Note that in general, Eq. (5.5), being considered as an equation for λ, may have a number of roots $\lambda = \lambda^{(j)}(x)$. Let us represent vector space W as a union $W = \bigcup_i D_i$, such that Eq. (5.5) can be resolved with respect to λ uniquely for any $x \in D_i$. In other words, in each domain D_i, the function x may serve as local coordinates on L_H as well as λ. This implies that the Lagrangian manifold L_H can be represented as $L_H = \bigcup_i L_i$, where

$$L_i = \left\{ x \in D_i \mid x = -\frac{\partial H}{\partial \lambda}, \right\}.$$

We shall call such domains L_i *phases*.

Thus, the results of the measurement of random vectors obtained by using the minimal information gain principle can be presented by means of either Legendrian submanifold in contact space $\tilde{L} \subset (\tilde{\Phi}, \tilde{\theta})$ or Lagrangian submanifold in symplectic space $L_H \subset (\Phi, \omega = -d\tilde{\theta})$, and all necessary functions ρ, $Z(\lambda)$, $H(\lambda)$, $I(p, q)$ can be directly derived from them.

5.2.4   Variance of Random Vectors

First of all, let us recall that the *second moment* of a random vector $X : (\Omega, \mathcal{A}, p) \to W$ is a symmetric 2-form $\mu_2(X) \in S^2(W)$

$$\mu_2(X) = \int_\Omega X(\omega) \otimes X(\omega) dp.$$

Central second moment or *variance* is a symmetric 2-form $\sigma_2(X) \in S^2(W)$

$$\sigma_2(X) = \mu_2(X - \mu_1(X)) = \mu_2(X) - \mu_1(X) \otimes \mu_1(X).$$

Let Hess(H) be the Hessian of the Hamiltonian H:

$$\text{Hess}(H) = \sum_{i,j} \frac{\partial^2 H}{\partial \lambda_i \partial \lambda_j} d\lambda_i \otimes d\lambda_j.$$

Theorem 5.2 *The variance $\sigma_2(X)$ of a random vector X is equal to $-$Hess(H):*

$$\sigma_2(X) = -\text{Hess}(H).$$

Let us define the differential quadratic form κ on Φ by the following way:

$$\kappa = \frac{1}{2} \sum_i (d\lambda_i \otimes dx_i + dx_i \otimes d\lambda_i) = d\lambda \cdot dx.$$

This differential quadratic form being restricted onto the manifold L_H takes the form

$$\kappa|_{L_H} = \frac{1}{2} \sum_i (d\lambda_i \otimes dx_i + dx_i \otimes d\lambda_i) \Bigg|_{\left\{ x = -\frac{\partial H}{\partial \lambda} \right\}} = -\text{Hess}(H) = \sigma_2(X).$$

Since the variance is non-negative, the differential quadratic form $\kappa|_{L_H}$ must be non-negative.

Thus, the manifold $\Phi = W \times W^*$ is equipped with two structures:

- symplectic structure

$$\omega = d\lambda \wedge dx,$$

- pseudo Riemannian structure of signature (n, n)

$$\kappa = d\lambda \cdot dx.$$

The measurement procedure of a random vector $X \colon (\Omega, \mathcal{A}, p) \to W$ is presented by the Lagrangian manifold $L_H \subset (\Phi, \omega)$ that has to be Riemannian manifold with respect to the quadratic differential form $\kappa|_{L_H}$. The last requirement leads us to the notion of *applicable phases*, i.e. Riemannian submanifolds of L_H, where both x and λ may serve as coordinates.

5.2.5 Thermodynamics

Here, we show that all above constructions allow us to consider thermodynamics as a measurement of extensive variables, such as energy, volume and mass.

First of all, one of the basic laws of thermodynamics, the energy conservation law, claims that the heat is consumed by the physical system for the changing of its internal energy and work, and particularly for gas-like systems, it has the form (here, we pretend that we already know the first part of the second law of thermodynamics $\delta Q = T dS$)

$$dS = T^{-1}dE + pT^{-1}dV - \gamma T^{-1}dm, \tag{5.11}$$

where S is the entropy, T is the temperature, p is the pressure, γ is the chemical potential, E is the energy, V is the volume and m is the mass.

It is absolutely clear from the physical point of view what is written in (5.11). But mathematically, we can see the identity of two 1-forms, which is possible iff $S = $ const, $V = $ const, $E = $ const and $m = $ const. Moreover, the second part of the second law of thermodynamics claims that

$$dS > 0$$

for irreversible processes, which means that we can compare differential 1-forms with zero.

All these issues of mathematical nature, together with a notion that (5.11) reminds us the similar contact structure appearing in measurement theory, drive us to consider thermodynamics as a theory of measuring extensive variables $(E, V, m) \in W$. The fact that W is a vector space corresponds to the additivity properties of extensives. Then, intensives $(-T^{-1}, -pT^{-1}, \gamma T^{-1}) \in W^*$ may serve as Lagrangian multipliers λ that we have seen in the above discussion. Once we put $n = 3$ and assign $(x_1, x_2, x_3) = (E, V, m)$ and $(\lambda_1, \lambda_2, \lambda_3) = (-T^{-1}, -pT^{-1}, \gamma T^{-1})$, we are able to reformulate the laws of thermodynamics in the following way:

- **The first law of thermodynamics**
  The state of any thermodynamical system described by intensives (p, T, γ), extensives (E, V, m) and entropy S is a Legendrian manifold $\tilde{L} \subset (\tilde{\Phi}, \tilde{\theta})$, where $\tilde{\Phi} = \mathbb{R} \times W \times W^*$, i.e. maximal integral manifold of the form

$$\tilde{\theta} = d(-S) - (-T^{-1})dE - (-pT^{-1})dV - (\gamma T^{-1})dm. \tag{5.12}$$

Comparing (5.12) with (5.9) and (5.10), we conclude that the following relation holds:

$$dI = -dS. \tag{5.13}$$

The postulate that the variance of random vectors is positive gives us what we will mean by the second thermodynamical law, and what in classical thermodynamics is usually called "conditions of thermodynamical stability". Below, in Sect. 5.3, we will show the explanation.

- **The second law of thermodynamics**

  The immersed Lagrangian manifold $L \subset (\Phi, \Omega = -d\tilde{\theta})$ obtained by restriction of natural projection $\pi : \tilde{\Phi} \to \Phi, \pi(S, E, V, m, p, T, \gamma) = (E, V, m, p, T, \gamma)$, which is a local diffeomorphism, onto the Legendrian manifold $\tilde{L}$ is equipped with the differential quadratic form

  $$\kappa = d(-T^{-1}) \cdot dE + d(-pT^{-1}) \cdot dV + d(\gamma T^{-1}) \cdot dm,$$

  and the only applicable domains on L are those ones where the form κ is positive.

From (5.13), one can conclude that the well-known entropy increasing law for irreversible processes (for example, the establishment of thermodynamical equilibrium between two systems) is the principle of minimal information gain from the measurement viewpoint.

From (5.13), it also follows that

$$S = -I + \alpha_0.$$

Since the information gain I is always greater than zero, the entropy has to be $S \leq \alpha_0$. This can be interpreted as the third law of thermodynamics.

5.3 Thermodynamics of Gases

Now that we have declared that thermodynamic states are Legendrian or Lagrangian surfaces in the corresponding contact or symplectic space, and we can give a more accurate description of gases in the form appropriate for further purposes, namely, the analysis of critical phenomena in filtration processes and Euler flows.

5.3.1 Specific Variables

First of all, we introduce the so-called *specific* thermodynamic variables by the following way. Let $S = S(E, V)$ be a function on the Legendrian surface $\tilde{L}$. Due to additive properties of the entropy S, the function $S(E, V)$ has to be homogeneous of degree 1 with respect to the mass of the system, i.e.

$$S(E, V, m) = mS\left(\frac{E}{m}, \frac{V}{m}\right).$$

Introduce the notation: $e = E/m$, $v = V/m$, $S\left(\frac{E}{m}, \frac{V}{m}\right) = s(e, v)$, and call e the specific energy, v the specific volume and $s(e, v)$ the specific entropy. Then, in terms of specific variables, the form $\tilde{\theta}$ can be written as

$$\tilde{\theta} = \left(-s + T^{-1}e + pT^{-1}v - \gamma T^{-1}\right)dm + \left(-ds + T^{-1}de + pT^{-1}dv\right)m.$$

If a thermodynamic state $\tilde{L}$ is now given by a function $s = s(e, v)$, then

$$\gamma = e - Ts + pv, \quad \left(-ds + T^{-1}de + pT^{-1}dv\right)\Big|_{\tilde{L}} = 0. \tag{5.14}$$

The differential quadratic form κ on L will take the form

$$\kappa = -m\left(d(T^{-1}) \cdot de + d(pT^{-1}) \cdot dv\right),$$

and since applicable domains are defined by the positivity of κ and mass m is assumed to be positive, the applicability condition is formulated as negativity of the form $-m^{-1}\kappa$, which we will continue, denoting by κ:

$$\kappa = d(T^{-1}) \cdot de + d(pT^{-1}) \cdot dv. \tag{5.15}$$

Since $G = E - TS + pV$ is the Gibbs free energy, $\gamma = e - Ts + pv$ is the specific Gibbs free energy.

5.3.2 Legendrian and Lagrangian Manifolds for Gases

Relation (5.14) allows us to define Legendrian surfaces by means of specific variables. Indeed, consider the contact space $(\mathbb{R}^5, \theta)$ equipped with coordinates (s, e, v, p, T) and contact 1-form

$$\theta = -ds + T^{-1}de + pT^{-1}dv.$$

Then, a thermodynamic state is a Legendrian manifold $\tilde{L}$, such that $\theta|_{\tilde{L}} = 0$. For a given function $s = s(e, v)$, this manifold is defined by relations:

$$s = s(e, v), \quad T = \frac{1}{s_e}, \quad p = \frac{s_v}{s_e}.$$

In practice, we actually do not have the function $s(e, v)$, because there are no devices that would allow us to measure the specific entropy $s(e, v)$. This brings us to the necessity of eliminating s from our consideration. To this end, we introduce the projection $\pi : \mathbb{R}^5 \rightarrow \mathbb{R}^4$, $\pi(s, e, v, p, T) = (e, v, p, T)$. Its restriction onto the Legendrian manifold $\tilde{L}$ is an immersed Lagrangian manifold $L = \pi(\tilde{L})$, while $\mathbb{R}^4$ is equipped with the symplectic form

$$\Omega = d\theta = d(T^{-1}) \wedge de + d(pT^{-1}) \wedge dv,$$

which vanishes on L: $\Omega|_L = 0$. The Lagrangian manifold $L \subset (\mathbb{R}^4, \Omega)$ is given by the two functions

$$f(e, v, p, T) = 0, \quad g(e, v, p, T) = 0. \tag{5.16}$$

The condition that L is Lagrangian is expressed by vanishing of the Poisson bracket $[f, g]$ on L, i.e. $[f, g]|_L = 0$, where

$$[f, g]\Omega \wedge \Omega = df \wedge dg \wedge \Omega.$$

This bracket in coordinates is of the form

$$[f, g] = \frac{1}{2} \left(pT \left(f_p g_e - f_e g_p \right) + T^2 \left(f_T g_e - f_e g_T \right) + T \left(f_v g_p - f_p g_v \right) \right).$$

In thermodynamics of gases, the functions in (5.16) usually have the form

$$f(e, v, p, T) = p - A(v, T), \quad g(e, v, p, T) = e - B(v, T). \tag{5.17}$$

The first equation of state in (5.17) is called *thermic*, and the second one is called *caloric*. From experiments, one can obtain the first state equation, but not the second one, because we have no devices that measure the specific energy. But having known the first equation and using the compatibility condition $[f, g]|_L = 0$, one gets the caloric equation, and therefore the Lagrangian surface for a given gas becomes completely determined. Then, relations

$$T = \frac{1}{s_e}, \quad p = \frac{s_v}{s_e} \tag{5.18}$$

can be considered as an overdetermined system for $s(e, v)$, which is compatible due to $[f, g]|_L = 0$. Solving (5.18), we get unknown function $s(e, v)$ and therefore define the Legendrian manifold $\tilde{L}$ completely.

Let us take the Poisson bracket between $f(e, v, p, T)$ and $g(e, v, p, T)$ in the form (5.17). Then, we get

$$(T^{-2}B)_v = (T^{-1}A)_T,$$

from what follows that the following theorem is valid.

Theorem 5.3 *The Lagrangian manifold L is given by the Massieu–Planck potential $\phi(v, T)$:*

$$p = RT\phi_v, \quad e = RT^2\phi_T, \tag{5.19}$$

and the specific entropy s and Gibbs free energy γ are

$$s = R(\phi + T\phi_T), \quad \gamma = RT(v\phi_v - \phi), \tag{5.20}$$

where $R = 8.314 \, J \cdot K^{-1} \cdot \text{mol}^{-1}$ is the universal gas constant.

Using (5.19), it is easy to show that the differential quadratic form (5.15) can be expressed as follows:

$$R^{-1}\kappa = - \left(\phi_{TT} + 2T^{-1}\phi_T\right) dT \cdot dT + \phi_{vv}dv \cdot dv.$$

Hence, applicable domains are defined by inequalities

$$\phi_{TT} + 2T^{-1}\phi_T > 0, \quad \phi_{vv} < 0. \tag{5.21}$$

From (5.21) and (5.19), it follows that applicable states are also given by

$$e_T > 0, \quad p_v < 0. \tag{5.22}$$

It is worth to say that relations in (5.22) are usually called *conditions of thermodynamical stability* with respect to thermic and mechanical perturbations, respectively.

5.3.3   Singularities of Lagrangian Manifolds and Phase Transitions

Let us now explore singularities of projections of the Lagrangian manifold L to spaces of intensive variables (p, T) and extensive variables (v, e). The singularities of the first type occur where the differential form $dp \wedge dT$ degenerates and coincides with the set where $\phi_{vv} = 0$. The singularities of the second type are the points where $de \wedge dv = 0$, or, equivalently, $\phi_{TT} + 2T^{-1}\phi_T = 0$. Thus, the set where the Lagrangian manifold L has singularities is exactly the set where the differential quadratic form κ degenerates. We will consider such Lagrangian manifolds that (e, v) serve as global coordinates on them. This means that L in our consideration will have no singularities of the second type, and since κ has to be negative, the inequality $\phi_{TT} + 2T^{-1}\phi_T > 0$ holds everywhere on L. Domains of the manifold L that have no singularities and on which the form κ is negative definite, we have already called *applicable phases*. Consequently, if κ is nondegenerate and negative on the entire manifold L, and therefore L has no singularities of its projections to the space of intensive variables as well, then a thermodynamical system corresponding to such a Lagrangian manifold L has only one phase, otherwise it has a number of phases, separated from one another by the set where κ is non-negative, or, where $\phi_{vv} \geq 0$. In the last case, thermodynamical system has a remarkable property called *phase transitions of the first type*.

Definition 5.1 A jump from one applicable point $(e_1, v_1) \in L$ to another applicable point $(e_2, v_2) \in L$, governed by the intensive variables (p, T) and the specific Gibbs potential γ conservation law, is called *phase transition of the first type*.

A set of points where phase transition occurs is a curve on the Lagrangian manifold L, which is called *coexistence* or *binodal* curve. Using (5.19) and (5.20), one gets the following equations for the coexistence curve $\Gamma \subset \mathbb{R}^3(p, v, T)$ [4]:

$$\phi_v(v_2, T) = \frac{p}{RT}, \quad \phi_v(v_1, T) = \frac{p}{RT}, \tag{5.23}$$

$$\phi(v_2, T) - \phi(v_1, T) - v_2\phi_v(v_2, T) + v_1\phi_v(v_1, T) = 0. \tag{5.24}$$

Thus, solving (5.23), we define the location for phases of thermodynamical system on the corresponding Lagrangian manifold L.

5.4 Examples of Gases

In this section, we discuss various models of real gases and show how above methods can be applied to the analysis of gases. We provide a detailed description for models of real gases, which are extremely important for applications—van der Waals, Peng–Robinson and Redlich–Kwong models.

5.4.1 Ideal Gases

We start with the simplest model of gases—ideal gases. The Lagrangian manifold L for ideal gases is given by equations

$$f(e, v, p, T) = p - Rv^{-1}T, \quad g(e, v, p, T) = e - \frac{nRT}{2}, \tag{5.25}$$

where n is the degree of freedom. The first equation in (5.25) is called *Mendeleev–Clapeyron* equation.

The Legendrian manifold $\tilde{L}$ is defined by (5.25) extended by

$$s = R \ln \left(T^{n/2} v \right) + \frac{Rn}{2}.$$

The Massieu–Planck potential coincides with the specific entropy s (up to a multiplicative constant R):

$$\phi = \ln \left(T^{n/2} v \right).$$

And finally, the differential quadratic form κ for ideal gases is

$$\kappa = -\frac{Rn}{2T^2} dT \cdot dT - R v^{-2} dv \cdot dv.$$

One can see that κ is negative; therefore, the Lagrangian manifold L for ideal gases has no singularities, and there are no phase transitions.

5.4.2   Van der Waals Gases

The van der Waals model is historically the first one admitting phase transitions of gas–liquid type. The thermic equation is of the form:

$$f(e, v, p, T) = p - \frac{RT}{v - b} + \frac{a}{v^2},$$

where a and b are constants responsible for the interaction between particles and their volume, respectively. Note that in case $a = 0$ and $b = 0$, one gets the ideal gas state equation.

To find out the second equation of state, we assume that $g(e, v, p, T) = e - B(v, T)$ and take the Poisson bracket $[f, g]$. Since it should be zero on the Lagrangian surface, we get the following equation for $B(v, T)$:

$$v^2 B_v - a = 0,$$

from what follows that $B(v, T) = F(T) - a/v$. Putting $a = 0$ and $b = 0$, we get an ideal gas, and the caloric equation for van der Waals gases is of the form

$$g(e, v, p, T) = e - \frac{nRT}{2} + \frac{a}{v}.$$

Let us now resolve system (5.18) for van der Waals gases. The result will be (up to additive constant)

$$s(v, T) = R \ln \left(T^{n/2}(v - b) \right) + \frac{Rn}{2}.$$

The Massieu–Planck potential has the following form:

$$\phi(v, T) = \ln \left(T^{n/2}(v - b) \right) + \frac{a}{vRT}.$$

Finally, the differential quadratic form κ for van der Waals gases is [1, 4]

$$\kappa = -\frac{Rn}{2T^2} dT \cdot dT - \frac{v^3 RT - 2a(v - b)^2}{v^3 T(v - b)^2} dv \cdot dv.$$

We can see that the first component of κ is negative, while the second one can change its sign. Therefore, the Lagrangian manifold has singularities of projections to the plane of intensive variables (p, T). The curve in coordinates (T, v) where the differential quadratic form κ changes its sign, which is also called *spinodal curve* given by

$$T = \frac{2a(v - b)^2}{Rv^3}.$$

This function has a maximum at point $v_{\text{crit}} = 3b$, and it equals $T_{\text{crit}} = 8a/27Rb$. The temperature T_{crit} is called *critical temperature*, and if $T > T_{\text{crit}}$, the differential quadratic form κ is negative. The corresponding critical values for pressure p_{crit}, energy e_{crit} and entropy s_{crit} could be found as well. It is more convenient to work with dimensionless thermodynamic variables. To this end, we introduce the following contact scale transformation:

$$T \longmapsto \frac{T}{T_{\text{crit}}}, \quad v \longmapsto \frac{v}{v_{\text{crit}}}, \quad p \longmapsto \frac{p}{p_{\text{crit}}}, \quad e \longmapsto \frac{e}{e_{\text{crit}}}, \quad s \longmapsto \frac{s}{s_{\text{crit}}},$$

where $T_{\text{crit}}, v_{\text{crit}}, p_{\text{crit}}, e_{\text{crit}}$ and s_{crit} are critical parameters for van der Waals gases:

$$T_{\text{crit}} = \frac{8a}{27Rb}, \quad v_{\text{crit}} = 3b, \quad p_{\text{crit}} = \frac{a}{27b^2}, \quad e_{\text{crit}} = \frac{a}{9b}, \quad s_{\text{crit}} = \frac{3R}{8};$$

then, we get the reduced equations of state in new dimensionless coordinates, which we will continue, denoting by p, T, e and v:

$$p = \frac{8T}{3v - 1} - \frac{3}{v^2}, \qquad e = \frac{4n}{3} T - \frac{3}{v}.$$

One can easily show that the Massieu–Planck potential and the specific entropy for van der Waals gases take the form:

$$\phi = \ln\left(T^{n/2}(3v-1)\right) + \frac{9}{8vT} + C_\phi, \quad s = \ln\left(T^{4n/3}(3v-1)^{8/3}\right) + C_s, \quad (5.26)$$

where constants C_ϕ and C_s are

$$C_\phi = \ln\left(\left(\frac{2}{3}\right)^{3n/2}\left(\frac{a}{bR}\right)^{n/2}b\right), \quad C_s = \frac{4n}{3} + \ln\left(\left(\frac{2}{3}\right)^{4n}\left(\frac{a}{bR}\right)^{4n/3}b^{8/3}\right).$$

And the differential quadratic form becomes

$$\kappa = -\frac{Rn}{2}\frac{dT^2}{T^2} - \frac{9R(4Tv^3 - 9v^2 + 6v - 1)}{4Tv^3(3v-1)^2}dv^2.$$

The spinodal curve together with the coexistence curve in coordinates (p, T) for van der Waals gases is presented in Fig. 5.1. The equations for the coexistence curve in reduced coordinates have the form [4]:

$$\frac{3p}{8T} = \frac{3}{3v_{1,2}-1} - \frac{9}{8v_{1,2}^2 T}, \quad \frac{(3v_1-1)(3v_2-1)(v_1+v_2)}{v_1-v_2}\ln\left(\frac{3v_2-1}{3v_1-1}\right)$$

$$= 3(v_1 + v_2 - 6v_1 v_2),$$

and the coexistence curve together with the spinodal curve for van der Waals gases in coordinates (p, v) is presented in Fig. 5.2.

Both curves on the Lagrangian manifold for van der Waals gases are shown in Fig. 5.3. The area on the left of the coexistence curve corresponds to the liquid phase, and the right area is the gas phase. The area inside the coexistence curve is a condensation of the gas, and the area between the coexistence and spinodal curves corresponds to possible thermodynamic states, but dramatically unstable. On the left, such states are called *overheated liquid*, while the right one is *overcooled gas*.

5.4.3 Peng–Robinson Gases

Another very important model of real gases is Peng–Robinson model proposed in [22]. It appeared to be effective in description of hydrocarbons. The first state equation has the following form:

$$f(p, T, e, v) = p - \frac{RT}{v-b} + \frac{a}{(v+b)^2 - 2b^2},$$

Fig. 5.1 Spinodal curve (red) and coexistence curve (blue) for van der Waals gases in coordinates (p, T)

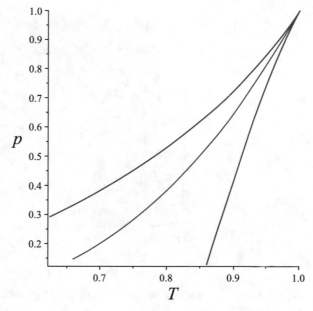

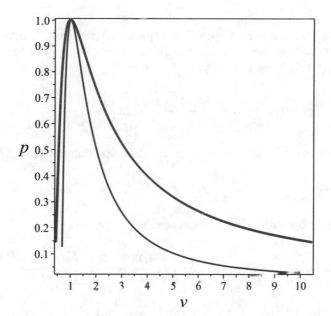

Fig. 5.2 Spinodal curve (red) and coexistence curve (blue) for van der Waals gases in coordinates (p, v)

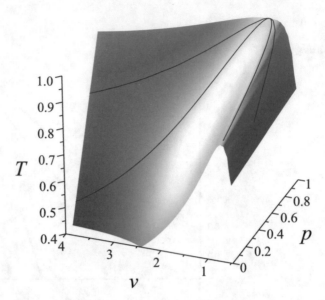

Fig. 5.3 Lagrangian manifold for van der Waals gases together with spinodal (red) and coexistence (blue) curves

where constants a and b are responsible for particles' interaction and their volume, respectively, as in the van der Waals model. The compatibility condition leads us to the following caloric equation of state:

$$g(p, T, e, v) = e - \frac{nRT}{2} - \frac{a\sqrt{2}}{4b} \ln\left(\frac{v + b - \sqrt{2}b}{v + b + \sqrt{2}b}\right).$$

As in case of van der Waals gases, let us introduce the contact scale transformation

$$p \longmapsto \frac{a}{b^2}p, \quad T \longmapsto \frac{a}{bR}T, \quad e \longmapsto \frac{a}{b}e, \quad v \longmapsto bv, \quad s \longmapsto Rs.$$

The reduced Peng–Robinson state equations are

$$p = \frac{T}{v - 1} - \frac{1}{(v + 1)^2 - 2}, \quad e = \frac{nT}{2} + \frac{\sqrt{2}}{4} \ln\left(\frac{v + 1 - \sqrt{2}}{v + 1 + \sqrt{2}}\right),$$

$$s = \ln\left(T^{n/2}(v - 1)\right) + \frac{n}{2} + \ln\left(\left(\frac{a}{bR}\right)^{n/2} b\right).$$

The Massieu–Planck potential ϕ for Peng–Robinson gases is

$$\phi(v, T) = \ln\left(T^{n/2}(v-1)\right) - \frac{\sqrt{2}}{4T}\ln\left(\frac{(3-2\sqrt{2})(v\sqrt{2}+v-1)}{v\sqrt{2}-v+1}\right) + \ln\left(\left(\frac{a}{bR}\right)^{n/2}b\right).$$

The differential quadratic form κ has the form

$$R^{-1}\kappa = -\frac{n}{2T^2}dT \cdot dT - \frac{Tv^4+2(2T-1)v^3+2(T+1)v^2-2(2T-1)v+T-2}{T(v-1)^2(v^2+2v-1)^2}dv \cdot dv.$$

Therefore, the singular set of the Lagrangian manifold for Peng–Robinson gases can be found from

$$T = \frac{2(v+1)(v-1)^2}{(v^2+2v-1)^2}.$$

As in case of van der Waals gases, there is a critical point $(T_{\text{crit}}, v_{\text{crit}})$, such that if $T > T_{\text{crit}}$, then the differential quadratic form κ is negative for any v.

Theorem 5.4 ([3]) *The critical temperature for Peng–Robinson gases T_{crit} and the corresponding critical volume v_{crit} are defined as follows:*

$$v_{\text{crit}} = 1 + 2(4+2\sqrt{2})^{-1/3} + (4+2\sqrt{2})^{1/3}, \quad T_{\text{crit}} = \frac{2(v_{\text{crit}}+1)(v_{\text{crit}}-1)^2}{(v_{\text{crit}}^2+2v_{\text{crit}}-1)^2}.$$

The coexistence curve for Peng–Robinson gases in coordinates (p, v, T) is presented in Fig. 5.4 and is of similar form as for van der Waals gases.

5.4.4 Redlich–Kwong Gases

The next model of real gases is the Redlich–Kwong model. It was proposed in [23] and became of wide popularity in filtration processes. The thermic equation of state for Redlich–Kwong gases is

$$f(p, T, v, e) = p - \frac{RT}{v-b} + \frac{a}{\sqrt{T}v(v+b)}. \tag{5.27}$$

If one takes the Poisson bracket $[f, g]|_L$, where $g(e, v, p, T) = e - B(v, T)$, one gets the following equation for $B(v, T)$:

$$3a - 2v\sqrt{T}(v+b)B_v = 0,$$

which has solution

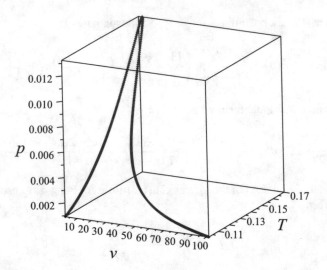

Fig. 5.4 Coexistence curve for Peng–Robinson gases

$$B(v, T) = F(T) + \frac{3a}{2b\sqrt{T}} \ln\left(\frac{v}{v+b}\right).$$

Again, $F(T)$ has to be defined as $F(T) = nRT/2$, and the Lagrangian manifold for Redlich–Kwong gases is given by (5.27) together with [2]

$$g(p, T, v, e) = e - \left(\frac{nRT}{2} + \frac{3a}{2b\sqrt{T}} \ln\left(\frac{v}{v+b}\right)\right). \tag{5.28}$$

Resolving the equation for the specific entropy s as described in Sect. 5.2, we get

$$s(v, T) = \frac{Rn}{2} \ln T + R \ln(v - b) + \frac{a}{2bT^{3/2}} \ln\left(\frac{v}{v+b}\right) + \frac{Rn}{2}. \tag{5.29}$$

Thus, the Legendrian manifold $\tilde{L}$ is defined by (5.27), (5.28) and (5.29).

The contact scale transformation

$$p \longmapsto \left(\frac{Ra^2}{b^5}\right)^{1/3} p, \quad T \longmapsto \left(\frac{a}{Rb}\right)^{2/3} T, \quad v \longmapsto bv,$$

$$e \longmapsto \left(\frac{Ra^2}{b^2}\right)^{1/3} e, \quad s \longmapsto Rs$$

gives us reduced state equations [2]:

$$p = \frac{T}{v - 1} - \frac{1}{\sqrt{T}v(v + 1)}, \quad e = \frac{nT}{2} + \frac{3}{2\sqrt{T}} \ln\left(\frac{v}{v+1}\right), \tag{5.30}$$

$$s = \ln\left(T^{n/2}(v-1)\right) + \frac{1}{2T^{3/2}} \ln\left(\frac{v}{v+1}\right) + \frac{n}{2} + \ln\left(\left(\frac{a}{Rb}\right)^{n/3} b\right). \quad (5.31)$$

The Massieu–Planck potential is of the form

$$\phi(v,T) = \ln\left(T^{n/2}(v-1)\right) - \frac{1}{T^{3/2}} \ln\left(\frac{v}{v+1}\right) + \ln\left(b\left(\frac{a}{Rb}\right)^{n/3}\right).$$

The differential quadratic form κ is

$$\kappa R^{-1} = -\left(\frac{n}{2T^2} + \frac{3}{4T^{7/2}} \ln(1+v^{-1})\right) dT \cdot dT$$

$$- \frac{v^2(v+1)^2 T^{3/2} - 2v^3 + 3v^2 - 1}{T^{3/2}(v+1)^2 v^2(v-1)^2} dv \cdot dv.$$

Note that in case of Redlich–Kwong gases, the first component of the differential quadratic form κ depends on the specific volume v. But since pressure p and temperature T are assumed to be positive, from (5.30), it follows that only $v > 1$ have sense, and therefore, the component mentioned is negative, and the projection of the Lagrangian manifold to (e, v) for Redlich–Kwong gases has no singularities.

Spinodal and coexistence curves for Redlich–Kwong gases can be elaborated in the same way as for van der Waals and Peng–Robinson gases and can be found in [2].

5.5   Basic Equations

Thermodynamics of real gases discussed in previous sections forms a base for the analysis of gas dynamics, since, as we will see further, equations describing dynamics significantly depend on thermodynamical properties. In this section, we formulate the problem and provide general solutions for stationary Darcy and Euler flows.

The system of equations describing one-component filtration of gases in porous media (Darcy flows) or inviscid gas flow (Euler flow) consists of [6–8]

1. The momentum conservation law

   • the Darcy law (for filtration processes)

$$\mathbf{u} = -\frac{k}{\mu} \nabla p, \quad (5.32)$$

where $\mathbf{u}(x) = (u_1, u_2, u_3)$ is the velocity field, $x \in D \subset \mathbb{R}^3$, $p(x)$ is the pressure, $k = k(v, T)$ and $\mu = \mu(v, T)$ are the coefficients of permeability and viscosity, respectively, which are the functions of the medium.

- the Euler equation (for inviscid gases)

$$\rho(\mathbf{u}, \nabla)\mathbf{u} = -\nabla p, \qquad (5.33)$$

where $\rho(x) = v^{-1}(x)$ is the density.

2. The mass conservation law

$$\operatorname{div}(\rho \mathbf{u}) = 0. \qquad (5.34)$$

In addition to (5.32), (5.33) and (5.34), we assume that the flows of both kinds are adiabatic, i.e. the specific entropy s is constant along the trajectories of the velocity field $\mathbf{u}$:

$$(\mathbf{u}, \nabla s) = 0. \qquad (5.35)$$

One can see that the system (5.32) or (5.33) together with (5.34) and (5.35) is incomplete. It is quite expectable, since we have not yet specified the medium above Eqs. (5.32)–(5.35) are written for. It can be done by means of equations of state (Legendrian manifold)

$$p = RT\phi_v, \quad e = RT^2\phi_T, \quad s = R(\phi + T\phi_T), \qquad (5.36)$$

where $\phi(v, T)$ is given.

Then, system of equations including (5.32) or (5.33), together with (5.34)–(5.36) is complete

Using equations of state together with equations describing dynamics not only makes the system complete but also allows us to investigate how thermodynamic properties, especially phase transitions, appear in solutions of equations. Indeed, having solution of basic equations, one gets thermodynamical variables as functions in D, and therefore coexistence curves for various gases can be moved from the Lagrangian manifold to D. This allows us to define the location for different phases of gases and the set of points in D where phase transition occurs.

Suppose that the domain D contains a number of sources located at points $a_i \in D$, having intensities J_i. We will denote the source as a pair $(a_i, J_i), i = \overline{1, N}$. Then, D can be represented as a union of domains $D = \cup D_k$, where each D_k contains sources with common specific entropy s_0, while filtrations in D_k are independent [4]. Therefore, we can restrict ourselves on case $s(x) = s_0$. We can also say that the gas is involved in an adiabatic process.

Geometrically, thermodynamical processes can be understood as contact transformations of $(\mathbb{R}^5, \theta)$ preserving the Legendrian manifold $\tilde{L}$, or, from infinitesimal viewpoint, as contact vector fields X tangent to $\tilde{L}$. Integral curve of X is a curve l on $\tilde{L}$, which we will mean by a thermodynamical process.

5.5.1 Darcy Flows

In case of Darcy flows, the following theorem is valid.

Theorem 5.5 *Let the thermodynamic state of the gas be given by $\tilde{L}$ and $l \subset \tilde{L}$ be a thermodynamical process. Then, filtration equations (5.32), (5.34) and (5.36) are equivalent to the Dirichlet problem*

$$\Delta(Q(\tau)) = 0, \quad \tau|_{\partial D} = \tau_0,$$

where

$$Q(\tau) = -\int v^{-1}(\tau) \frac{k(\tau)}{\mu(\tau)} p'(\tau) d\tau,$$

τ is a parameter on l and Δ is the Laplace operator.

Proof Let τ be a parameter on a given process l. Then, all the thermodynamic variables can be expressed in terms of τ, in particular,

$$p = p(\tau), \quad v = v(\tau), \quad T = T(\tau), \quad k = k(\tau), \quad \mu = \mu(\tau). \tag{5.37}$$

From (5.32) and (5.34), one gets

$$0 = \operatorname{div}(v^{-1}\mathbf{u}) = \operatorname{div}\left(v^{-1}(\tau)\left(-\frac{k(\tau)}{\mu(\tau)}\nabla p\right)\right) = \operatorname{div}\left(v^{-1}(\tau)\left(-\frac{k(\tau)}{\mu(\tau)}p'(\tau)\nabla\tau\right)\right)$$

$$= \operatorname{div}\left(Q'(\tau)\nabla\tau\right) = \operatorname{div}\left(\nabla Q(\tau)\right) = \Delta(Q(\tau)),$$

where $Q(\tau) = -\int v^{-1}(\tau)\frac{k(\tau)}{\mu(\tau)}p'(\tau)d\tau$. $\square$

This theorem is a generalization of that in [1–4]. The result of this theorem gives an explicit method of finding solutions for the Dirichlet filtration problem. Note that this result is of general form for all gases and all processes. All we need is to find the function $Q(\tau)$ for a given gas and a given process.

In case of N sources (a_i, J_i) and $D = \mathbb{R}^3$, one has

$$\tau(x) = Q^{-1}\left(\sum_{i=1}^{N} \frac{J_i}{4\pi|x - a_i|} + Q(\tau_0)\right).$$

By means of (5.37), we have $p = p(x)$, $v = v(x)$, $T = T(x)$ and so on and therefore get a complete solution for Darcy flows. The conditions for the invertibility of $Q(\tau)$ will be formulated for concrete gases.

5.5.2 Euler Flows

For Euler flows, we analyze in detail the case of one point isotropic source (a, J) in D and have the same theorem as for Darcy flows [5].

Theorem 5.6 *Let the thermodynamic state of the gas be given by $\tilde{L}$ and $l \subset \tilde{L}$ be a thermodynamical process. Then, the solution for problems (5.33), (5.34) and (5.36) is given implicitly by the following formula:*

$$\frac{v^2(\tau)}{2|x - a|^4} + \left(\frac{4\pi}{J}\right)^2 \Psi(\tau) = 0, \tag{5.38}$$

where τ is a parameter on l and

$$\Psi(\tau) = \int v(\tau) p'(\tau) d\tau.$$

Proof Let $r = |x - a|$ be a distance from the source, $x - a = \mathbf{r}$, and let $\mathbf{n} = \mathbf{r}/r$. Since the source is isotropic, one has

$$\mathbf{u} = U(r)\mathbf{r}, \quad \nabla = \mathbf{n}\partial_r.$$

The intensity of the source is equal to J, which means that the mass flux through a sphere S_a of radius r with a centre at a is equal to J:

$$J = \int_{S_a} v^{-1}(\tau(r))(\mathbf{u}, \mathbf{n}) dS = 4\pi r^3 v^{-1}(\tau(r)) U(r),$$

from what follows that

$$U(r) = \frac{J}{4\pi r^3} v(\tau(r)).$$

Then, Eq. (5.33) due to (5.37) takes the form

$$v^{-1}(\tau) \left(\frac{J}{4\pi}\right)^2 \frac{v(\tau)}{r^2} \frac{d}{dr} \left(\frac{v(\tau)}{r^2}\right) = -p'(\tau) \frac{d\tau}{dr},$$

which in turn becomes

$$\frac{d}{dr}\left(\frac{v^2(\tau)}{2r^4}\right) + \left(\frac{4\pi}{J}\right)^2 v(\tau)p'(\tau)\frac{d\tau}{dr} = 0,$$

from what follows the statement of the theorem. □

Once one computes the function $\Psi(\tau)$, one gets a complete solution for the stationary Euler problem.

In our case, the specific volume v can be chosen as a parameter τ on adiabatic process $l_{\text{adiab}} \subset \tilde{L}$. Indeed, due to (5.20), we have the following relation:

$$s_0 = R(\phi + T\phi_T),$$

which can be considered as an equation for $T(v)$ since the derivative of the right-hand side is positive in an applicable domain. Therefore, all the thermodynamical variables can be expressed in terms of v.

5.6 Solutions

In this section, we discuss solutions of Darcy and Euler equations for concrete gases using the analysis of their thermodynamic properties in Sect. 5.4. For simplicity, the permeability coefficient k and the viscosity μ are assumed to be constants.

5.6.1 Darcy Flows

5.6.1.1 Ideal Gases

First of all, let us express all the thermodynamic variables in terms of v. For ideal gases, we have

$$T(v) = \exp\left(\frac{2s_0}{Rn}\right)v^{-2/n}, \quad p(v) = R\exp\left(\frac{2s_0}{Rn}\right)v^{-2/n-1}. \tag{5.39}$$

Therefore, the function $Q(v)$ equals

$$Q(v) = -\frac{Rk}{2\mu}\exp\left(\frac{2s_0}{Rn}\right)\frac{n+2}{n+1}v^{-2/n-2}.$$

And finally, the solution $v(x)$ has the following form:

$$v(x) = \left(Q(v_0) - \frac{2\mu}{Rk} \exp\left(-\frac{2s_0}{Rn}\right) \frac{n+1}{n+2} \sum_{i=1}^{N} \frac{J_i}{4\pi |x - a_i|} \right)^{-\frac{n}{2(n+1)}},$$

where v_0 is the specific volume at infinity.

5.6.1.2 van der Waals Gases

For van der Waals gases, the expressions for $T(v)$ and $p(v)$ are of the form:

$$T(v) = \exp\left(\frac{3s_0}{4n}\right)(3v-1)^{-1-2/n}, \quad p(v) = 8\exp\left(\frac{3s_0}{4n}\right)(3v-1)^{-1-2/n} - \frac{3}{v^2}.$$

And the function $Q(v)$ is defined by the relation [4]

$$-\frac{\mu}{k}Q(v) = -\frac{2}{v^3} + 8\exp\left(\frac{3s_0}{4n}\right)\frac{(3v-1)^{-\alpha}}{v} + 8\exp\left(\frac{3s_0}{4n}\right)\int (3v-1)^{-\alpha}v^{-2}dv,$$

where $\alpha = 1 + 2/n$.

For van der Waals gases, the conditions for invertibility of $Q(v)$ are given by the following theorem [1, 4].

Theorem 5.7 *The function $Q(v)$ is invertible if the specific entropy constant s_0 satisfies the following inequality:*

$$\exp\left(\frac{3s_0}{4n}\right) > \frac{1}{4\alpha}(1+\alpha)^{1+\alpha}(2-\alpha)^{2-\alpha}.$$

Thus, if the above condition holds, the solution is uniquely determined, otherwise, there are a number of possibilities in filtration development. The case of multivalued solution for one source is considered in detail in [4]. Here, we concentrate on a uniquely determined solution for a number of sources. As we have said, having solution for the filtration problem, one can find the location of different phases. It is presented in Fig. 5.5. We can see that the condensation process is observed in the neighbourhood of the sources.

5.6.1.3 Peng–Robinson Gases

In case of Peng–Robinson gases, the expressions for $T(v)$ and $p(v)$ have the following form:

$$T(v) = \exp\left(\frac{2s_0}{n}\right)(v-1)^{-2/n}, \quad p(v) = \exp\left(\frac{2s_0}{n}\right)(v-1)^{-1-2/n} - \frac{1}{(v+1)^2 - 2}.$$

Fig. 5.5 Distribution of
phases for van der Waals
gases. Coloured domain
corresponds to the
condensation process, and
white domain corresponds to
gas phase

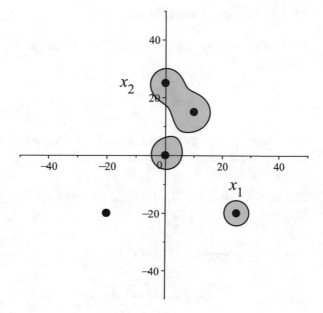

Therefore, the function $Q(v)$ is

$$-\frac{\mu}{k}Q(v) = \frac{3\sqrt{2}}{4}\ln\left(\frac{v+1+\sqrt{2}}{v+1-\sqrt{2}}\right) + \ln\left(\frac{v^2}{v^2+2v-1}\right) -$$

$$-\frac{v+2}{v^2+2v-1} - \exp\left(\frac{2s_0}{n}\right)\left(1+\frac{2}{n}\right)\int\frac{dv}{v(v-1)^{2+2/n}}.$$

Invertibility conditions for $Q(v)$ in this case can be given by the following
theorem [3].

Theorem 5.8 *The function $Q(v)$ is invertible if the specific entropy constant s_0
satisfies the inequality:*

$$\exp\left(\frac{2s_0}{n}\right) > \frac{2n(v_0+1)(v_0-1)^{2+2/n}}{(n+2)(v_0+2v_0-1)^2},$$

where v_0 is the root of the equation:

$$(2-n)v^3 + 3(n+2)v^2 + (3n+2)v + 3n - 2 = 0. \tag{5.40}$$

There exists a real root of (5.40), $v_0 > 1$.

The distribution for phases has the same form as in case of van der Waals gases and
can be found in [3].

5.6.1.4 Redlich–Kwong Gases

For Redlich–Kwong gases, the given level of the specific entropy s_0 leads us to the following relation:

$$s_0 = \ln\left(T^{n/2}(v-1)\right) + \frac{1}{2T^{3/2}}\ln\left(\frac{v}{v+1}\right). \tag{5.41}$$

We cannot get an explicit expression for $T(v)$ from (5.41), but nevertheless, one can estimate the asymptotic behaviour for $p(v)$, $T(v)$ and $Q(v)$ when $v \to 1$ or $v \to \infty$ [2].

Theorem 5.9 *If $v \to 1$, then asymptotics for $T(v)$, $p(v)$ and $Q(v)$ have the following form:*

$$T(v) = \frac{B^{2/3}}{(v-1)^{2/3}} + O\left((v-1)^{1/3}\right), \quad p(v) = \frac{B^{2/3}}{(v-1)^{5/3}} + O\left(\frac{1}{(v-1)^{2/3}}\right),$$

$$Q(v) = -\frac{kB^{2/3}}{\mu(v-1)^{5/3}} + O\left(\frac{1}{(v-1)^{2/3}}\right),$$

where $B = \exp(s_0)$.

Theorem 5.10 *If $v \to +\infty$, then asymptotics for $T(v)$, $p(v)$ and $Q(v)$ have the following form:*

$$T(v) = \frac{1}{(B^*v)^{2/3}} + O\left(\frac{1}{v^{5/3}}\right), \quad p(v) = \frac{c}{v^{5/3}} + O\left(\frac{1}{v^{8/3}}\right),$$

$$Q(v) = -\frac{5kc}{8\mu v^{8/3}} + O\left(\frac{1}{v^{11/3}}\right),$$

where B^* is the root of the equation

$$-s_0 = B/2 + \ln B,$$

and

$$c = \left(B^*\right)^{-2/3} - \left(B^*\right)^{1/3}.$$

Let us now analyze the invertibility conditions for $Q(v)$. We need to find actually the conditions for s_0 when $Q(v)$ is monotonic for all v that have sense, i.e. $v > 1$. In other words, $Q'(v)$ should have no zeroes for $v > 1$. But since the relations $Q'(v) = 0$ and $p'(v) = 0$ are equivalent, one has to explore $p'(v)$ including s_0 as a parameter. First of all, using equation of state $p(T, v)$, one has

$$\frac{dp}{dv} = \frac{\partial p}{\partial v} + \frac{\partial p}{\partial T}\frac{dT}{dv}. \tag{5.42}$$

The derivative $T'(v)$ can be obtained by means of (5.41). Once we substitute it in (5.42), the equation $p'(v) = 0$ will take the form

$$AZ^2 + BZ + C = 0, \tag{5.43}$$

where $Z = T^{3/2}$ and

$$A = 2(n+2)v^2(v+1)^2, \quad C = (v-1)^2\left(1 - 3(2v+1)\ln\left(1+v^{-1}\right)\right),$$

$$B = 3v^2(v+1)^2\ln\left(1+v^{-1}\right) + 2(v-1)((v+1)(2v+n) - 2nv^2).$$

Since $A > 0$ and $C < 0$, the discriminant of (5.43) is positive, and therefore, there are two real roots. But since $B > 0$, one of them is negative and is out of consideration. Thus, from equation $p'(v) = 0$, we have

$$T(v) = \left(\frac{-B + \sqrt{B^2 - 4AC}}{2A}\right)^{2/3}. \tag{5.44}$$

Substituting the root (5.44) in (5.41), we get the relation

$$s_0 = H(v). \tag{5.45}$$

If the specific entropy level s_0 is such that (5.45) has no solution $v^* > 1$, then $Q(v)$ is invertible. An example for $H(v)$ in case of $n = 3$ is presented in Fig. 5.6. Numerical computation shows that if $s_0 > -0.5$, then $Q(v)$ is invertible [2]. The distribution of phases is very similar to the case of van der Waals gases and can be found in [2].

5.6.2 Euler Flows

Here, we discuss the solution for Euler flows of ideal and van der Waals gases. Peng–Robinson and Redlich–Kwong models can be elaborated in the same way.

First of all, we take $D = \mathbb{R}^3$ assuming that the specific volume is given at infinity $v|_{|x-a|\to\infty} = v_0$. Since we take v as a parameter on the process l_{adiab}, the general formula (5.38) takes the form [5]:

$$\frac{v^2}{2|x-a|^4} + \left(\frac{4\pi}{J}\right)^2 \Psi(v) = 0,$$

where $\Psi(v) = \int vp'(v)dv$.

Fig. 5.6 Graph of function $H(v)$

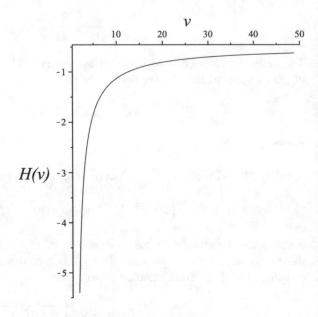

5.6.2.1 Ideal Gases

Using (5.39), one can show that the function $\Psi(v)$ for ideal gases has the following form:

$$\Psi(v) = \frac{R(n+2)}{2} \exp\left(\frac{2s_0}{Rn}\right) v^{-2/n}.$$

Therefore, the solution for ideal gases has the form (it is more convenient to work in terms of density $\rho = v^{-1}$ here):

$$\frac{1}{2|x-a|^4\rho^2} + \left(\frac{4\pi}{J}\right)^2 \exp\left(\frac{2s_0}{Rn}\right) \frac{R(n+2)}{2}\rho^{2/n} = C_0,$$

where C_0 is a constant depending on $\rho|_{|x-a|\to\infty} = \rho_0$.

Theorem 5.11 ([5]) *If $\rho_0 = 0$, then the asymptotic behaviour of $\rho(x)$ at infinity is of the form:*

$$\rho(x) = \frac{1}{\sqrt{2C_0}|x-a|^2} + o\left(\frac{1}{|x-a|^2}\right),$$

Fig. 5.7 The distribution of
the density for ideal gases

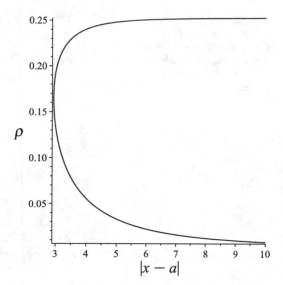

and if $\rho_0 \neq 0$, then

$$\rho(x) = \left(\frac{J}{4\pi}\right)^n \left(\frac{2C_0}{R(n+2)} \exp\left(-\frac{2s_0}{Rn}\right)\right)^{n/2} + \sum_{i=1}^{\infty} \frac{\beta_i}{|x-a|^{4i}}.$$

Thus, the solution obtained is regular at infinity in both cases. The distribution of the density is shown in Fig. 5.7.

We can see that the solution is multivalued. Moreover, it exists not for any x [5].

Theorem 5.12 *The solution $\rho(x)$ exists if*

$$|x-a| > \left(2\rho_*^2\left(C_0 - R\left(\frac{4\pi}{J}\right)^2 \exp\left(\frac{2s_0}{Rn}\right)(n/2+1)\rho_*^{2/n}\right)\right)^{-1/4},$$

where

$$\rho_* = \left(\frac{J}{4\pi}\right)^n \left(\exp\left(-\frac{2s_0}{Rn}\right)\frac{2nC_0}{R(n+1)(n+2)}\right)^{n/2}.$$

5.6.2.2   van der Waals Gases

In case of van der Waals gases, the solution is given by the following formula [5]:

$$\frac{1}{2|x-a|^4\rho^2} + \left(\frac{4\pi}{J}\right)^2 \times$$

$$\left(4\exp\left(\frac{3s_0}{4n}\right)(3\rho^{-1}-1)^{-(1+2/n)}\left(\rho^{-1}(n+2)-\frac{n}{3}\right)-6\rho\right) = C_0.$$

Fig. 5.8 The distribution of phases for van der Waals gases. Variable $y = 1$ is the liquid phase, and variable $y = 0.5$ is the condensation process

Fig. 5.9 The distribution of phases for van der Waals gases. Variable $y = 0$ is the gas phase, and variable $y = 0.5$ is the condensation process

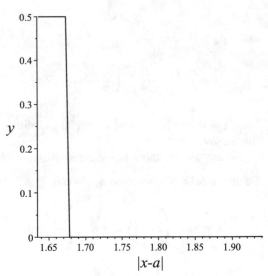

We analyze phase transitions in two cases, for "small" and "big" levels of the specific entropy constant s_0. Let us start with the higher branch of solution.

If $s_0 = 0.5$, the distribution of phases is shown in Fig. 5.8. We observe the condensation process near the source, while far from the source, the medium is in a liquid phase.

If $s_0 = 200$, the distribution of phases is shown in Fig. 5.9. We again observe the condensation process near the source, but at large distances from the source, the medium is in a gas phase.

For the lower branch of solution, in both cases, the gas condensates.

5.7 Conclusions

In this paper, we presented the analysis of critical phenomena in two types of stationary gas flows—filtration flows described by the Darcy law and flows of inviscid gases described by Euler equations. We showed that thermodynamics plays a crucial role in modelling of gas dynamics, and since taking into account thermodynamic properties of the medium expressed by state equations, we not only make the system of continuous media equations complete but also get an opportunity to investigate how these properties influence the flow. We provided constructive methods of finding solutions for Darcy and Euler flows, which are of general form not only for various gases but also for thermodynamical processes these gases are involved in. We also showed that the solutions of both Darcy and Euler system are, in general, multivalued, and for Darcy flows, the conditions for uniquely determined solution can be formulated, while in case of Euler flows, the solution is always multivalued, but each branch is determined by the conditions at infinity, and therefore only one of them can be realized physically. Another important property of solutions is their regularity at infinity, usually accepted in physics as "correctness". The analysis of critical phenomena showed that Darcy and Euler flows have the same distribution of phases in case of "big level" of the specific entropy.

Further investigations in this field can be connected with the analysis of critical phenomena in case of, in some sense, distributed sources, i.e. the source is no longer assumed to be a point but occupies some domain. Another direction could be, on the one hand, the exploration of gas flows on Riemannian manifolds, and on the other hand, the investigation of flows for media with more complicated thermodynamics, namely, phase transitions of the higher order, which requires in turn the more detailed analysis of geometric structures on the corresponding Legendrian and Lagrangian manifolds.

Acknowledgments This work was partially supported by the Russian Foundation for Basic Research (project 18-29-10013). The authors are also grateful to the anonymous reviewer whose suggestions improved the paper, and special thanks for the possibility to include Fig. 5.3 into the paper.

References

1. Lychagin, V.: Adiabatic filtration of an ideal gas in a homogeneous and isotropic porous medium. Glob. Stoch. Anal. 6(1), 23–31 (2019)
2. Lychagin, V., Roop, M.: Phase transitions in filtration of Redlich-Kwong gases. J. Geom. Phys. **143**, 33–40 (2019)
3. Lychagin, V., Roop, M.: Steady filtration of Peng-Robinson gases in a porous medium. Glob. Stoch. Anal. 6(2), 59–68 (2019)
4. Lychagin, V., Roop, M.: Phase transitions in filtration of real gases. arXiv:1903.00276

5. Lychagin, V., Roop, M.: Real gas flows issued from a source. Anal. Math. Phys. 10(1), Article: 3 (2019). https://doi.org/10.1007/s13324-019-00349-z
6. Scheidegger, A.E.: The physics of flow through porous media. Revised edition. The Macmillan Co., New-York (1960)
7. Leibenson, L.S.: Motion of natural liquids and gases in a porous medium. Gostekhizdat, Moscow (1947)
8. Muskat, M.: The Flow of Homogeneous Fluids Through Porous Media. McGraw-Hill, New-York (1937)
9. Maikov, I.L., Zaichenko, V.M., Torchinsky, V.M.: Theoretical Investigations of Filtration Process of Hydrocarbon Mixtures in Porous Media. 9th International Conference on Heat Transfer, Fluid Mechanics and Thermodynamics, 1266–1269 (2012)
10. Kachalov, V.V, Molchanov, D.A., Sokotushchenko, V.N., Zaichenko, V.M.: Mathematical modeling of gas-condensate mixture filtration in porous media taking into account non-equilibrium of phase transitions. Journal of Physics: Conference Series. **774**(1) (2016)
11. Duyunova, A., Lychagin, V.,Tychkov, S.: Non-stationary adiabatic filtration of gases in porous media. arXiv:1908.09316
12. Landau, L.: New Exact Solution of the Navier-Stokes Equations. Doklady Akademii Nauk SSSR. **43** 311–314 (1944) (in Russian)
13. Broman, G.I., Rudenko, O.: Submerged Landau jet: exact solutions, their meaning and application. Uspekhi Fizicheskikh Nauk. **80**(1), 97–104 (2010) (in Russian)
14. Vinogradov, A.M., Krasilshchik, I.S.: Symmetries and Conservation Laws for Differential Equations of Mathematical Physics. Factorial, Moscow (1997)
15. Krasilshchik, I.S., Lychagin, V.V., Vinogradov, A.M.: Geometry of Jet Spaces and Nonlinear Partial Differential Equations. Gordon and Breach Science Publishers (1986)
16. Squire, H.B.: The round laminar jet. Quarterly Journal of Mechanics and Applied Mathematics. **4**(3), 321–329 (1951)
17. Roop, M.D.: Singular Solutions in Nonlinear Models of Fluid Dynamics. Proceedings of the 11th International Conference "Management of Large-Scale System Development" (MLSD) (2018). https://doi.org/10.1109/MLSD.2018.8551782
18. Gorinov, A., Lychagin, V., Roop, M., Tychkov, S.: Gas Flow with Phase Transitions: Thermodynamics and the Navier-Stokes Equations. In: Kycia, R., Ulan, M., Schneider, E. (eds.) Nonlinear PDEs, Their Geometry, and Applications., pp. 229–241. Birkhauser, Cham (2019)
19. Lychagin, V.: Contact Geometry, Measurement and Thermodynamics. In: Kycia, R., Ulan, M., Schneider, E. (eds.) Nonlinear PDEs, Their Geometry, and Applications., pp. 3–52. Birkhauser, Cham (2019)
20. Mrugala, R.: Geometrical formulation of equilibrium phenomenological thermodynamics. Reports on Mathematical Physics. **14**(3), 419–427 (1978)
21. Ruppeiner, G.: Riemannian geometry in thermodynamic fluctuation theory. Reviews of Modern Physics. **67**(3), 605–659 (1978)
22. Peng, B.Y., Robinson, D.B.: A New Two-Constant Equation of State. Industrial and Engineering Chemistry: Fundamentals. **15** 59–64 (1976)
23. Redlich, O., Kwong, J.N.S.: On the Thermodynamics of Solutions. Chem. Rev. **44**(1), 233–244 (1949).

Chapter 6
Differential Invariants for Flows of Fluids and Gases

Anna Duyunova, Valentin V. Lychagin, and Sergey Tychkov

6.1 Introduction

The paper is an extended overview of the papers [10–16]. The main extension is a detailed analysis of thermodynamic states, symmetries, and differential invariants. This analysis is based on consideration of Riemannian structure [7] naturally associated with Lagrangian manifolds that represent thermodynamic states. This approach radically changes the description of the thermodynamic part of the symmetry algebra as well as the field of differential invariants.

The paper is organized as follows.

In Sect. 6.2 we discuss thermodynamics in terms of contact and symplectic geometries. The main part of this approach is a presentation of thermodynamic states as Lagrangian manifolds equipped with an additional Riemannian structure. Application of this approach to fluid motion is new, though the relationship between contact geometry and thermodynamics was well-known since Gibbs [2] and Carathéodory [3]. For some modern studies see also [4] and [5].

In Sect. 6.3 the motion of inviscid media is considered. We discuss flows of inviscid fluids on different Riemannian manifolds: a plane, sphere, and a spherical layer. Such flows are governed by a generalization of the Euler equation system. For each of these cases, we find a Lie algebra of symmetries, provide a classification of symmetry algebras depending on a thermodynamic state admitted by media, and describe the field of differential invariants for the Euler system.

In Sect. 6.4 the motion of viscid media on Riemannian manifolds is studied. First, we discuss a generalization of the Navier–Stokes equations for an arbitrary oriented Riemannian manifold. Then for the cases of a plane, space, sphere, and

A. Duyunova · V. V. Lychagin · S. Tychkov (✉)
Institute of Control Sciences of Russian Academy of Sciences, Moscow, Russia
e-mail: valentin.lychagin@uit.no; sergey.lab06@ya.ru

© The Author(s), under exclusive license to Springer Nature Switzerland AG 2021   187
M. Ulan, E. Schneider (eds.), *Differential Geometry, Differential Equations, and Mathematical Physics*, Tutorials, Schools, and Workshops in the Mathematical Sciences, https://doi.org/10.1007/978-3-030-63253-3_6

a spherical layer, we provide classification of symmetry algebras with respect to possible thermodynamic states and give full description for the field of differential invariants.

6.2   Thermodynamics

Here we consider the media with thermodynamics described by two types of quantities. The first are *extensive* quantities: the specific entropy s, the specific volume ρ^{-1}, the specific internal energy ϵ; and the second are *intensive* quantities: the absolute temperature $T > 0$ and the pressure p.

A thermodynamic state of such media is a two-dimensional Legendrian manifold $L \subset \mathbb{R}^5(\epsilon, \rho, p, T, s)$, a maximal integral manifold of the differential 1-form

$$\theta = d\epsilon - T ds - p\rho^{-2}d\rho,$$

i.e. a manifold such that the first law of thermodynamics $\theta\big|_L = 0$ holds.

Following [7] a point $(\epsilon, \rho, p, T, s)$ on the Legendrian manifold can be considered as a triplet: the expected value (ϵ, ρ^{-1}) of a stochastic process of measurement of internal energy and volume, the probabilistic measure corresponding to (p, T) and the information $(-s)$, which is given up to a constant.

Since the information I is a positive quantity, the entropy s satisfies the inequality $s \leqslant s_0$ for a certain constant s_0, which, generally speaking, depends on the nature of a process under consideration.

Let us denote the variance of the stochastic process by $(-\kappa)$. In terms of the given probabilistic measure and expected values it has the form [7]:

$$\kappa = d(T^{-1}) \cdot d\epsilon - \rho^{-2}d(pT^{-1}) \cdot d\rho.$$

Thus, by a **thermodynamic state** we mean a two-dimensional Legendrian submanifold L of the contact manifold $(\mathbb{R}^5, \theta)$, such that the quadratic differential form κ on the surface L is negative definite, i.e.

$$\kappa\big|_L < 0.$$

Because the energy can be excluded from the conservation laws that govern medium motion, we also eliminate it from our geometrical interpretation of the thermodynamics.

Consider the projection

$$\varphi : \mathbb{R}^5 \to \mathbb{R}^4, \quad \varphi : (\epsilon, \rho, p, T, s) \longmapsto (\rho, p, T, s).$$

The restriction of the map φ on the state surface L is a local diffeomorphism on the image $\tilde{L} = \varphi(L)$ and the surface $\tilde{L}$ is an immersed Lagrangian manifold in the symplectic space $\mathbb{R}^4$ equipped with the structure form

$$\Omega = ds \wedge dT + \rho^{-2} d\rho \wedge dp.$$

Therefore, the first law of thermodynamics is equivalent to the condition that $\tilde{L} \subset \mathbb{R}^4$ is a Lagrangian manifold.

The two-dimensional surface $\tilde{L}$ in the four-dimensional space can be defined by two equations

$$f(p, \rho, s, T) = 0, \quad g(p, \rho, s, T) = 0 \tag{6.1}$$

and the condition for the surface $\tilde{L}$ to be Lagrangian means vanishing of the Poisson bracket of these functions

$$[f, g]\big|_{\tilde{L}} = 0, \tag{6.2}$$

that in the coordinates (p, ρ, s, T) takes the form

$$[f, g] = \rho^2 \left(f_\rho g_p - f_p g_\rho \right) + f_s g_T - f_T g_s.$$

Thus, *the thermodynamic state* can be defined as Lagrangian surface (6.1) in the four-dimensional symplectic space, such that the condition (6.2) holds and the symmetric differential form κ is negative definite on this surface.

Note, if the equation of state is given in the form $\epsilon = \epsilon(\rho, s)$, then the two-dimensional Legendrian manifold L can be defined by the structure equations

$$\epsilon = \epsilon(\rho, s), \quad T = \epsilon_s, \quad p = \rho^2 \epsilon_\rho, \tag{6.3}$$

and the restriction of the form κ gives

$$\kappa\big|_L = -\epsilon_s^{-1} \left(\left(\epsilon_{\rho\rho} + 2\rho^{-1} \epsilon_\rho \right) d\rho^2 + 2\epsilon_{\rho s} d\rho \cdot ds + \epsilon_{ss} ds^2 \right).$$

The condition of negative definiteness for this form leads us to the following additional relations

$$\begin{cases} \epsilon_{\rho\rho} + 2\rho^{-3} p > 0, \\ \epsilon_{ss} \left(\epsilon_{\rho\rho} + 2\rho^{-3} p \right) - \epsilon_{\rho s}^2 > 0 \end{cases}$$

on the function $\epsilon(\rho, s)$ or

$$\begin{cases} p_\rho > 0, \\ T_s p_\rho - \rho^2 T_\rho^2 > 0. \end{cases}$$

6.3 Compressible Inviscid Fluids or Gases

In this section we study differential invariants of compressible inviscid fluids or gases.

The system of differential equations (the Euler system) describing flows on an oriented Riemannian manifold (M, g) consists of the following equations (see [1] for details):

$$\begin{cases} \rho(\mathbf{u}_t + \nabla_{\mathbf{u}}\mathbf{u}) = -\operatorname{grad} p + \mathbf{g}\rho, \\[2mm] \dfrac{\partial(\rho\,\Omega_g)}{\partial t} + \mathcal{L}_{\mathbf{u}}\left(\rho\,\Omega_g\right) = 0, \\[2mm] T\,(s_t + \nabla_{\mathbf{u}}s) - \dfrac{k}{\rho}\Delta_g T = 0, \end{cases} \tag{6.4}$$

where the vector field $\mathbf{u}$ is the flow velocity, p, ρ, s, T are the pressure, density, entropy, temperature of the fluid respectively, k is the thermal conductivity, which is supposed to be constant, and $\mathbf{g}$ is the gravitational acceleration.

Here ∇_X is the directional covariant Levi–Civita derivative with respect to a vector field X, $\mathcal{L}_X$ is the Lie derivative along a vector field X, Ω_g is the volume form on the manifold M, Δ_g is the Laplace–Beltrami operator corresponding to the metric g.

The first equation of system (6.4) represents the law of momentum conservation in the inviscid medium, the second is the continuity equation, and the third is the equation representing the effect of heat conduction in the medium.

We consider the following examples of manifold M: a plane, sphere and a spherical layer.

Note that in all these cases the number of unknown functions is greater than the number of system equations by 2, i.e. the system (6.4) is incomplete. We get two additional equations taking into account the thermodynamics of the medium.

Thus, by the Euler system of differential equations we mean the system (6.4) extended by two equations of state (6.1), such that the functions f and g satisfy the additional relation (6.2) and the form κ is negative definite.

Geometrically, we represent this system in the following way. Consider the bundle of rank $(\dim M + 4)$

$$\pi : \mathbb{R} \times TM \times \mathbb{R}^4 \longrightarrow \mathbb{R} \times M,$$

where $(t, \mathbf{x}, \mathbf{u}, \rho, p, T, s) \to (t, \bar{x})$ and $t \in \mathbb{R}$, $\mathbf{x} \in M$, $\mathbf{u} \in T_{\mathbf{x}}M$. Then the Euler system is a system of differential equations on sections of the bundle π.

Note that system (6.1) defines the zeroth order system $\mathcal{E}_0 \subset J^0\pi$.

Denote by $\mathcal{E}_1 \subset J^1\pi$ the system of order $\leqslant 1$ obtained by the first prolongation of the system $\mathcal{E}_0$ and by the first 2 equations of system (6.4) (Euler's and the continuity equations).

Let also $\mathcal{E}_2 \subset J^2\pi$ be the system of differential equations of order $\leqslant 2$ obtained by the first prolongation of the system $\mathcal{E}_1$ and the last equation of system (6.4).

For the case $k \geqslant 3$, we define $\mathcal{E}_k \subset J^k\pi$ to be the $(k-2)$-th prolongation of the system $\mathcal{E}_2$.

Note that due to the relations (6.1) the system $\mathcal{E}_\infty = \varprojlim \mathcal{E}_k$ is a formally integrable system of differential equations, which we also call the Euler system.

6.3.1   2D-Flows

Consider Euler system (6.4) on a plane $M = \mathbb{R}^2$ equipped with the coordinates (x, y) and the standard flat metric $g = dx^2 + dy^2$.

The velocity field of the flow has the form $\mathbf{u} = u(t, x, y)\,\partial_x + v(t, x, y)\,\partial_y$, the pressure p, the density ρ, the temperature T and the entropy s are the functions of time and space with the coordinates (t, x, y).

Here we consider the flow without any external force field, so $\mathbf{g} = 0$.

6.3.1.1   Symmetry Lie Algebra

The symmetry algebra of the Euler system has been found in [10], here we observe the main statements.

First of all, by a symmetry of the PDE system we mean a point symmetry, i.e. a vector field X on the jet space $J^0\pi$ such that its second prolongation $X^{(2)}$ is tangent to the submanifold $\mathcal{E}_2 \subset J^2\pi$.

To describe the Lie algebra of symmetries of the Euler system, we consider the Lie algebra $\mathfrak{g}$ generated by the following vector fields on the space $J^0\pi$:

$$X_1 = \partial_x, \qquad\qquad\qquad X_4 = t\,\partial_x + \partial_u,$$
$$X_2 = \partial_y, \qquad\qquad\qquad X_5 = t\,\partial_y + \partial_v,$$
$$X_3 = y\,\partial_x - x\,\partial_y + v\,\partial_u - u\,\partial_v, \qquad X_6 = \partial_t,$$
$$X_7 = \partial_s, \qquad X_{10} = t\,\partial_t + x\,\partial_x + y\,\partial_y - s\,\partial_s,$$
$$X_8 = \partial_p, \qquad X_{11} = t\,\partial_t - u\,\partial_u - v\,\partial_v - 2p\,\partial_p + s\,\partial_s,$$
$$X_9 = T\partial_T, \qquad X_{12} = p\,\partial_p + \rho\,\partial_\rho - s\,\partial_s.$$

Note that this symmetry algebra consists of pure geometric and thermodynamic parts.

The geometric part $\mathfrak{g}_m$ is generated by the fields $X_1, \ldots, X_6$. Transformations corresponding to the elements of this algebra are generated by the motions, Galilean transformations and the time shift.

In order to describe the pure thermodynamic part of the system symmetry algebra, consider the Lie algebra $\mathfrak{h}$ generated by the vector fields

$$Y_1 = \partial_s, \qquad Y_3 = \rho \, \partial_\rho, \qquad Y_5 = p \, \partial_p,$$

$$Y_2 = \partial_p, \qquad Y_4 = s \, \partial_s, \qquad Y_6 = T \, \partial_T.$$

Denote by $\vartheta : \mathfrak{g} \longmapsto \mathfrak{h}$ the following Lie algebras homomorphism:

$$\vartheta(X) = X(\rho)\partial_\rho + X(s)\partial_s + X(p)\partial_p + X(T)\partial_T, \tag{6.5}$$

where $X \in \mathfrak{g}$.

Note that, the kernel of the homomorphism ϑ is the ideal $\mathfrak{g}_m \subset \mathfrak{g}$.

Let also $\mathfrak{h}_t$ be the Lie subalgebra of the algebra $\mathfrak{h}$ that preserves the thermodynamic state (6.1).

Theorem 6.1 ([10]) *A Lie algebra $\mathfrak{g}_{\mathfrak{sym}}$ of symmetries of the Euler system of differential equations on a plane coincides with*

$$\vartheta^{-1}(\mathfrak{h}_t).$$

Note that, for the general equation of state, the algebra $\mathfrak{h}_t = 0$, and the symmetry algebra coincides with the Lie algebra $\mathfrak{g}_m$.

Observe that, usually, the equations of state are neglected, and vector fields like $f(t) \, \partial_p$ and $g(t)T \, \partial_T$, where f and g are arbitrary functions, are considered as symmetries of the Euler system.

6.3.1.2 Symmetry Classification of States

In this section we classify the thermodynamic states or the Lagrangian surfaces $\tilde{L}$ depending on the dimension of the symmetry algebra $\mathfrak{h}_t \subset \mathfrak{h}$.

We consider one- and two-dimensional symmetry algebras only, because the requirement on the thermodynamic state to have a three or more dimensional symmetry algebra is very strict and leads to the trivial solutions.

States with a One-Dimensional Symmetry Algebra

Let $\dim \mathfrak{h}_t = 1$ and let $Z = \sum_{i=1}^{6} \lambda_i Y_i$ be a basis vector in this algebra.

The state $\tilde{L} \subset \mathbb{R}^4$ is Lagrangian, i.e. $\Omega|_{\tilde{L}} = 0$, and therefore the vector field Z is tangent to the surface $\tilde{L}$, if and only if the differential 1-form

$$\iota_Z \Omega = \frac{\lambda_3}{\rho} dp - \frac{\lambda_5 p + \lambda_2}{\rho^2} d\rho - \lambda_6 T \, ds + (\lambda_4 s + \lambda_1) \, dT$$

vanishes on the surface $\tilde{L}$.

In other words, the surface $\tilde{L}$ is the solution of the following system of differential equations:

$$\begin{cases} \Omega|_{\tilde{L}} = 0, \\ (\iota_Z \Omega)|_{\tilde{L}} = 0. \end{cases}$$

In terms of specific energy (6.3) the last system has the following form:

$$\begin{cases} \lambda_3 \rho \, \epsilon_{\rho\rho} + (\lambda_4 s + \lambda_1)\epsilon_{\rho s} + (2\lambda_3 - \lambda_5) \, \epsilon_\rho - \dfrac{\lambda_2}{\rho^2} = 0, \\ (\lambda_4 s + \lambda_1)\epsilon_{ss} + \lambda_3 \rho \, \epsilon_{\rho s} - \lambda_6 \, \epsilon_s = 0. \end{cases}$$

It is easy to check that the bracket of these last two equations (see [8]) vanishes, and therefore the system is formally integrable and compatible.

In order to solve the last system we reduce its order and get the equivalent system

$$\begin{cases} \lambda_3 \rho \, \epsilon_\rho + (\lambda_4 s + \lambda_1)\epsilon_s + (\lambda_3 - \lambda_5)\epsilon + \dfrac{\lambda_2}{\rho} + f(s) = 0, \\ \lambda_3 \rho \, \epsilon_\rho + (\lambda_4 s + \lambda_1)\epsilon_s - (\lambda_6 + \lambda_4)\epsilon + g(\rho) = 0, \end{cases}$$

where $f(s)$ and $g(\rho)$ are some differentiable functions.

Below we list solutions of the system under the assumption of parameters λ generality. The more detailed description can be found in [10].

In the general case, when $\lambda_6 + \lambda_4 - \lambda_5 + \lambda_3 \neq 0$, solving the last system we find

$$p = C_1 \rho^{\frac{\lambda_5}{\lambda_3}} - \frac{\lambda_2}{\lambda_5}, \quad T = C_2(\lambda_4 s + \lambda_1)^{\frac{\lambda_6}{\lambda_4}},$$

where C_1, C_2 are constants.

Moreover, the negative definiteness of the quadratic differential form κ on the surface $\tilde{L}$ leads to the relations

$$\frac{\lambda_4 s + \lambda_1}{\lambda_6} > 0, \quad \frac{C_1 \lambda_5}{\lambda_3} > 0$$

for all $s \in (-\infty, s_0]$.

Theorem 6.2 *The thermodynamic states admitting a one-dimensional symmetry algebra have the form*

$$p = C_1 \rho^{\frac{\lambda_5}{\lambda_3}} - \frac{\lambda_2}{\lambda_5}, \quad T = C_2 (\lambda_4 s + \lambda_1)^{\frac{\lambda_6}{\lambda_4}},$$

where the constants defining the symmetry algebra satisfy inequalities

$$s_0 < -\frac{\lambda_1}{\lambda_4}, \quad C_1 > 0, \quad \frac{\lambda_5}{\lambda_3} > 0, \quad \frac{\lambda_2}{\lambda_5} < 0,$$

and besides they must meet one of the following conditions:

1. if $\frac{\lambda_6}{\lambda_4}$ is irrational, then $\lambda_4 < 0$, $\lambda_6 > 0$, $C_2 > 0$;
2. if $\frac{\lambda_6}{\lambda_4}$ is rational, then $\frac{\lambda_6}{\lambda_4} < 0$ (i.e. $\frac{\lambda_6}{\lambda_4} = -\frac{m}{k}$) and

   a. if k is even, then $\lambda_4 < 0$, $C_2 > 0$;
   b. if k is odd and m is even, then $C_2 > 0$;
   c. if k is odd and m is odd, then $C_2 \lambda_4 < 0$.

States with a Two-Dimensional Non-commutative Symmetry Algebra

Let $\mathfrak{h}_t \subset \mathfrak{h}$ be a non-commutative two-dimensional Lie subalgebra. Then $[\mathfrak{h}, \mathfrak{h}] \supset [\mathfrak{h}_t, \mathfrak{h}_t] = \langle Y_1, Y_2 \rangle$. Therefore, any non-zero vector $A = \alpha_0 Y_1 + \beta_0 Y_2 \in \mathfrak{h}_t$ can be chosen as one of the basis vectors. The second basis vector B in the subalgebra may be chosen such that $[A, B] = A$. Let $B = \sum_{i=1}^{6} \gamma_i Y_i$, then the condition $[A, B] = A$ gives two relations

$$\alpha_0 (\gamma_4 - 1) = 0, \quad \beta_0 (\gamma_5 - 1) = 0.$$

Restriction of the forms $\iota_A \Omega$ and $\iota_B \Omega$ on the surface L leads us to the following system of differential equations:

$$\begin{cases} \alpha_0 \epsilon_{ss} = 0, \\ \alpha_0 \rho^2 \epsilon_{\rho s} - \beta_0 = 0, \\ \gamma_3 \rho \epsilon_{\rho s} + (\gamma_4 s + \gamma_1) \epsilon_{ss} - \gamma_6 \epsilon_s = 0, \\ \gamma_3 \left(\rho \epsilon_{\rho\rho} + 2\epsilon_\rho \right) + (\gamma_4 s + \gamma_1) \epsilon_{\rho s} - \gamma_5 \epsilon_{\partial\rho} - \frac{\gamma_2}{\rho^2} = 0. \end{cases}$$

Note that from the first two equations of this system follows that $\alpha_0 \neq 0$. Computing brackets [8] we get that this system is integrable if

$$\beta_0 (\gamma_4 - \gamma_5) = 0, \quad \beta_0 (\gamma_3 \gamma_5 + \gamma_4 \gamma_6) = 0.$$

Then solving this system for the case $\beta_0 = 0$ and $\gamma_4 = 1$ we have $T = 0$ which is not sensible from the physical point of view.

For the case $\gamma_4 = 1$ and $\gamma_5 = 1$ we get

$$p = C\rho^{\frac{1}{\gamma_3}} + \frac{\beta_0}{\alpha_0}(s + \gamma_1) - \gamma_2, \quad T = -\frac{\beta_0}{\alpha_0 \rho},$$

but the condition on the form κ gives

$$\frac{C}{\gamma_3} > 0, \quad -\frac{1}{\rho^2} > 0.$$

So there are no thermodynamic states that admit a two-dimensional non-commutative symmetry algebra.

States with a Two-Dimensional Commutative Symmetry Algebra

Let now $\mathfrak{h}_{\mathfrak{t}} \subset \mathfrak{h}$ be a commutative two-dimensional Lie subalgebra, and let $A = \sum_{i=1}^{6} \alpha_i Y_i$, $B = \sum_{i=1}^{6} \beta_i Y_i$ be basis vectors in the algebra $\mathfrak{h}_{\mathfrak{t}}$.

Then the condition $[A, B] = 0$ gives the following relations on α's and β's:

$$\alpha_1\beta_4 - \alpha_4\beta_1 = 0, \quad \alpha_2\beta_5 - \alpha_5\beta_2 = 0. \tag{6.6}$$

Then, as above, restriction of the forms $\iota_A \Omega$ and $\iota_B \Omega$ on the state surface $\tilde{L}$ leads us to the following system of differential equations:

$$\begin{cases} \alpha_3\rho\,\epsilon_{\rho\rho} + (\alpha_4 s + \alpha_1)\epsilon_{\rho s} + (2\alpha_3 - \alpha_5)\,\epsilon_\rho - \dfrac{\alpha_2}{\rho^2} = 0, \\[2mm] \beta_3\rho\,\epsilon_{\rho\rho} + (\beta_4 s + \beta_1)\epsilon_{\rho s} + (2\beta_3 - \beta_5)\,\epsilon_\rho - \dfrac{\beta_2}{\rho^2} = 0, \\[2mm] (\alpha_4 s + \alpha_1)\epsilon_{ss} + \alpha_3\rho\,\epsilon_{\rho s} - \alpha_6\,\epsilon_s = 0, \\[2mm] (\beta_4 s + \beta_1)\epsilon_{ss} + \beta_3\rho\,\epsilon_{\rho s} - \beta_6\,\epsilon_s = 0. \end{cases}$$

The formal integrability condition for this system has the form

$$(\beta_5 - 5\beta_3)(\alpha_2\beta_5 - \alpha_5\beta_2) = 0,$$

which is satisfied due to relations (6.6).

Therefore, this system is integrable, and for all α's and β's. In most of cases this system has the "nonphysical" solution of the form $\epsilon = C_1\rho^{-1} + C_2$. For the special case, for example,

$$\alpha_1 = \frac{\alpha_4\beta_1}{\beta_4}, \quad \alpha_2 = \frac{\alpha_5\beta_2}{\beta_5} \quad \text{and} \quad \begin{cases} \alpha_3 = \alpha_5 - \alpha_4 - \alpha_6, \\ \beta_3 = \beta_5 - \beta_4 - \beta_6 \end{cases} \tag{6.7}$$

we have the following expressions for the pressure and the temperature:

$$p = C_1\rho^{\frac{\varsigma_2}{\varsigma_1+\varsigma_2}}(\beta_4 s + \beta_1)^{\frac{\varsigma_3+\varsigma_2}{\varsigma_1+\varsigma_2}} - \frac{\beta_2}{\beta_5}, \quad T = C_2\rho^{\frac{\varsigma_2}{\varsigma_1+\varsigma_2}}(\beta_4 s + \beta_1)^{\frac{\varsigma_3+\varsigma_2}{\varsigma_1+\varsigma_2}}\rho^{-1}(\beta_4 s + \beta_1)^{-1},$$

where

$$\varsigma_1 = \alpha_6\beta_4 - \alpha_4\beta_6, \quad \varsigma_2 = \alpha_4\beta_5 - \alpha_5\beta_4, \quad \varsigma_3 = \alpha_6\beta_5 - \alpha_5\beta_6.$$

And negative definiteness of the form κ leads to the relations

$$\frac{\varsigma_2(\beta_4 s + \beta_1)}{\varsigma_3\beta_4} > 0, \quad \frac{-\varsigma_1}{(\varsigma_1 + \varsigma_2)(\varsigma_2 + \varsigma_3)} > 0.$$

Theorem 6.3 *In the general case, there are no physically applicable thermodynamic states, which admit a two-dimensional commutative symmetry algebra.*

For the special case (6.7), the thermodynamic states admitting a two-dimensional commutative symmetry algebra have the form

$$p = C_1\rho^{\frac{\varsigma_2}{\varsigma_1+\varsigma_2}}(\beta_4 s + \beta_1)^{\frac{\varsigma_3+\varsigma_2}{\varsigma_1+\varsigma_2}} - \frac{\beta_2}{\beta_5}, \quad T = C_2\rho^{\frac{-\varsigma_1}{\varsigma_1+\varsigma_2}}(\beta_4 s + \beta_1)^{\frac{\varsigma_3-\varsigma_1}{\varsigma_1+\varsigma_2}},$$

where the constants defining the symmetry algebra satisfy inequalities

$$s_0 < -\frac{\beta_1}{\beta_4}, \quad \frac{\beta_2}{\beta_5} < 0, \quad \frac{\varsigma_2}{\varsigma_3} < 0, \quad \frac{\varsigma_1}{(\varsigma_1 + \varsigma_2)(\varsigma_2 + \varsigma_3)} < 0,$$

and besides they must meet one of the conditions:

1. *if $\frac{\varsigma_3+\varsigma_2}{\varsigma_1+\varsigma_2}$ is irrational, then $\beta_4 < 0$, $C_1 > 0$, $C_2 > 0$;*
2. *if $\frac{\varsigma_3+\varsigma_2}{\varsigma_1+\varsigma_2}$ is rational (i.e. $\frac{\varsigma_3+\varsigma_2}{\varsigma_1+\varsigma_2} = \pm\frac{m}{k}$), then*

   a. *if k is even, then $\beta_4 < 0$, $C_1 > 0$, $C_2 > 0$;*
   b. *if k is odd and m is even, then $C_1\beta_4 < 0$, $C_2 > 0$;*
   c. *if k is odd and m is odd, then $C_2\beta_4 < 0$, $C_1 > 0$.*

6.3.1.3 Differential Invariants

We consider two group actions on the Euler equation $\mathcal{E}$. The first one is the prolonged action of the group generated by the action of the Lie algebra $\mathfrak{g}_m$. The second action is the action generated by the prolongation of the action of the Lie algebra $\mathfrak{g}_{s\eta m}$.

First of all, observe that fibers of the projection $\mathcal{E}_k \to \mathcal{E}_0$ are irreducible algebraic manifolds.

Then we say that a function J on the manifold $\mathcal{E}_k$ is a *kinematic differential invariant of order $\leqslant k$* if

1. J is a rational function along fibers of the projection $\pi_{k,0} : \mathcal{E}_k \to \mathcal{E}_0$,
2. J is invariant with respect to the prolonged action of the Lie algebra $\mathfrak{g}_m$, i.e.

$$X^{(k)}(J) = 0, \tag{6.8}$$

for all $X \in \mathfrak{g}_m$.

Here we denote by $X^{(k)}$ the k-th prolongation of a vector field $X \in \mathfrak{g}_m$.

We say also that the kinematic invariant is *an Euler invariant* if condition (6.8) holds for all $X \in \mathfrak{g}_{s\eta m}$.

We say that a point $x_k \in \mathcal{E}_k$ and the corresponding orbit $\mathcal{O}(x_k)$ ($\mathfrak{g}_m$ or $\mathfrak{g}_{s\eta m}$-orbit) are *regular*, if there are exactly $m = \text{codim}\,\mathcal{O}(x_k)$ independent invariants (kinematic or Euler) in a neighborhood of this orbit.

Thus, the corresponding point on the quotient $\mathcal{E}_k/\mathfrak{g}_m$ or $\mathcal{E}_k/\mathfrak{g}_{s\eta m}$ is smooth, and these independent invariants (kinematic or Euler) can serve as local coordinates in a neighborhood of this point.

Otherwise, we say that the point and the corresponding orbit are *singular*.

It is worth to note that the Euler system together with the symmetry algebras $\mathfrak{g}_m$ or $\mathfrak{g}_{s\eta m}$ satisfies the conditions of Lie–Tresse theorem (see [9]), and therefore the kinematic and Euler differential invariants separate regular $\mathfrak{g}_m$ and $\mathfrak{g}_{s\eta m}$ orbits on the Euler system $\mathcal{E}$ correspondingly.

By a $\mathfrak{g}_m$ or $\mathfrak{g}_{s\eta m}$-invariant derivation we mean a total derivation

$$\nabla = A\frac{d}{dt} + B\frac{d}{dx} + C\frac{d}{dy}$$

that commutes with prolonged action of algebra $\mathfrak{g}_m$ or $\mathfrak{g}_{s\eta m}$. Here A, B, C are rational functions on the prolonged equation $\mathcal{E}_k$ for some $k \geqslant 0$.

The Field of Kinematic Invariants

First of all, observe that the functions

$$\rho, \quad s$$

(as well as p and T) on the equation $\mathcal{E}_0$ are $\mathfrak{g}_m$-invariants.

Straightforward computations using DifferentialGeometry package by I. Anderson [6] in Maple show that the following functions are the first order kinematic invariants:

$$J_1 = u_x + v_y, \qquad J_5 = \rho_x s_y - \rho_y s_x,$$

$$J_2 = u_y - v_x, \qquad J_6 = s_t + s_x u + s_y v,$$

$$J_3 = \rho_x^2 + \rho_y^2, \qquad J_7 = \rho_x(\rho_x u_x + \rho_y u_y) + \rho_y(\rho_x v_x + \rho_y v_y),$$

$$J_4 = s_x^2 + s_y^2, \qquad J_8 = s_x(\rho_x u_x + \rho_y u_y) + s_y(\rho_x v_x + \rho_y v_y).$$

It is easy to check that the codimension of the regular $\mathfrak{g}_m$-orbits on $\mathcal{E}_1$ is equal to 10.

Proposition 6.1 *The singular points belong to the union of two sets:*

$$\Upsilon_1 = \{ u_x - v_y = 0, \ u_y + v_x = 0, \ u_t = v_t = \rho_x = \rho_y = s_x = s_y = 0 \},$$

$$\Upsilon_2 = \{ J_3 J_5 (J_3 J_4 - J_5^2) = 0 \}.$$

The set Υ_1 contains singular points that have five-dimensional singular orbits. The set Υ_2 contains points where differential invariants $J_1, J_2, \ldots, J_8$ are dependent.

The proofs of the following theorems can be found in [10].

Theorem 6.4 ([10]) *The field of the first order kinematic invariants is generated by the invariants $\rho, s, J_1, J_2, J_3, \ldots, J_8$. These invariants separate the regular $\mathfrak{g}_m$-orbits.*

Theorem 6.5 ([10]) *The derivations*

$$\nabla_1 = \frac{d}{dt} + u \frac{d}{dx} + v \frac{d}{dy}, \quad \nabla_2 = \rho_x \frac{d}{dx} + \rho_y \frac{d}{dy}, \quad \nabla_3 = s_x \frac{d}{dx} + s_y \frac{d}{dy}$$

are $\mathfrak{g}_m$-invariant. They are linearly independent if

$$\rho_x s_y - \rho_y s_x \neq 0.$$

The bundle $\pi_{2,1} : \mathcal{E}_2 \to \mathcal{E}_1$ has rank 14, and by applying the derivations $\nabla_1, \nabla_2, \nabla_3$ to the kinematic invariants $J_1, J_2, \ldots, J_8$ we get 24 kinematic invariants. Straightforward computations show that among these invariants 14 are always independent (see https://d-omega.org).

Moreover, starting with the order $k = 1$ dimensions of regular orbits are equal to $\dim \mathfrak{g}_m = 6$ and all equations $\mathcal{E}_k, k \geq 3$, are the prolongations of $\mathcal{E}_2$.

Therefore, if we denote by $H(k)$ the Hilbert function of the $\mathfrak{g}_m$-invariants field, i.e. $H(k)$ is the number of independent invariants of pure order k (see [9] for details), then $H(k) = 5k + 4$ for $k \geq 2$, and $H(0) = 2, H(1) = 8$.

The corresponding Poincaré function is equal to

$$P(z) = \frac{2 + 4z - z^3}{(1 - z)^2}.$$

Summarizing, we get the following result.

Theorem 6.6 ([10]) *The field of the kinematic invariants is generated by the invariants ρ, s of order zero, by the invariants $J_1, J_2, \ldots, J_8$ of order one, and by the invariant derivations $\nabla_1, \nabla_2, \nabla_3$. This field separates the regular orbits.*

The Field of Euler Invariants

Let us consider the case when the thermodynamic state admits a one-dimensional symmetry algebra generated by the vector field

$$A = \xi_1 X_7 + \xi_2 X_8 + \xi_3 X_9 + \xi_4 X_{10} + \xi_5 X_{11} + \xi_6 X_{12}.$$

Note that, the $\mathfrak{g}_m$-invariant derivations $\nabla_1, \nabla_2, \nabla_3$ do not commute with the thermodynamic symmetry A.

Moreover, the action of the thermodynamic vector field A on the field of kinematic invariants is given by the following derivation:

$$\xi_6 \rho \partial_\rho + (\xi_1 - s(\xi_4 - \xi_5 + \xi_6)) \partial_s - J_1(\xi_4 + \xi_5)\partial_{J_1} - J_2(\xi_4 + \xi_5)\partial_{J_2} -$$

$$2J_3(\xi_4 - \xi_6)\partial_{J_3} - 2J_4(2\xi_4 - \xi_5 + \xi_6)\partial_{J_4} - J_5(3\xi_4 - \xi_5)\partial_{J_5} - J_6(2\xi_4 + \xi_6)\partial_{J_6} -$$

$$J_7(3\xi_4 + \xi_5 - 2\xi_6)\partial_{J_7} - 4\xi_4 J_8 \partial_{J_8}.$$

Therefore, finding the first integrals of this vector field we get the basic Euler invariants of the first order.

Theorem 6.7 ([10]) *The field of the Euler differential invariants for thermodynamic states admitting a one-dimensional symmetry algebra is generated by the differential invariants*

$$\frac{J_1}{J_6}\left(s - \frac{\xi_1}{\xi_4 - \xi_5 + \xi_6}\right), \quad J_1 \rho^{\frac{\xi_4 + \xi_5}{\xi_6}}, \quad \frac{J_2}{J_1}, \quad \frac{J_3}{\rho^3 J_6},$$

$$\frac{J_4 J_1^2}{\rho J_6^3}, \quad \frac{J_5 J_1}{J_8}, \quad J_6 \rho^{\frac{2\xi_4}{\xi_6}+1}, \quad \frac{J_7}{J_1 J_3}, \quad \frac{J_8}{\rho^2 J_6^2}$$

of the first order and by the invariant derivations

$$\rho^{\frac{\xi_4 + \xi_5}{\xi_6}} \nabla_1, \quad \rho^{\frac{2\xi_4}{\xi_6}-1} \nabla_2, \quad \rho^{\frac{3\xi_4 - \xi_5}{\xi_6}+1} \nabla_3.$$

This field separates the regular orbits.

Now consider the case when the thermodynamic state admits a commutative two-dimensional symmetry algebra generated by the vector fields $A = \sum\limits_{i=1}^{6} \mu_i X_{i+6}$ and $B = \sum\limits_{i=1}^{6} \eta_i X_{i+6}$ such that μ's and η's satisfy relations

$$\begin{cases} \eta_1\mu_4 - \eta_4\mu_1 - \eta_1\mu_5 + \eta_5\mu_1 + \eta_1\mu_6 - \eta_6\mu_1 = 0, \\ 2\eta_2\mu_5 - 2\eta_5\mu_2 - \eta_2\mu_6 + \eta_6\mu_2 = 0. \end{cases}$$

Using similar computations we get the following result.

Theorem 6.8 ([10]) *The field of the Euler differential invariants for thermodynamic states admitting a commutative two-dimensional symmetry algebra is generated by differential invariants*

$$J_1\rho^{\frac{\varsigma_1+\varsigma_2-2\varsigma_3}{\varsigma_2-\varsigma_1}}((\mu_4-\mu_5+\mu_6)s-\mu_1)^{\frac{\varsigma_2+\varsigma_1}{\varsigma_2-\varsigma_1}}, \quad \frac{J_2}{J_1}, \quad \frac{J_3}{\rho^3 J_1((\mu_4-\mu_5+\mu_6)s-\mu_1)},$$

$$\frac{\rho^8 J_1^2 J_4}{J_3^3}, \quad \frac{\rho^4 J_1 J_5}{J_3^2}, \quad \frac{\rho^3 J_6}{J_3}, \quad \frac{\rho^4 J_7}{J_3^2}, \quad \frac{J_8}{J_1 J_3}$$

of the first order and by the invariant derivations

$$\rho^{\frac{\varsigma_1+\varsigma_2-2\varsigma_3}{\varsigma_2-\varsigma_1}}((\mu_4-\mu_5+\mu_6)s-\mu_1)^{\frac{\varsigma_2+\varsigma_1}{\varsigma_2-\varsigma_1}}\nabla_1, \quad \rho^{\frac{3\varsigma_1-\varsigma_2-2\varsigma_3}{\varsigma_2-\varsigma_1}}((\mu_4-\mu_5+\mu_6)s-\mu_1)^{\frac{2\varsigma_1}{\varsigma_2-\varsigma_1}}\nabla_2,$$

$$\rho^{\frac{2\varsigma_1-2\varsigma_3}{\varsigma_2-\varsigma_1}}((\mu_4-\mu_5+\mu_6)s-\mu_1)^{\frac{3\varsigma_1-\varsigma_2}{\varsigma_2-\varsigma_1}}\nabla_3,$$

where

$$\varsigma_1 = \eta_4\mu_6 - \eta_6\mu_4, \quad \varsigma_2 = \eta_5\mu_6 - \eta_6\mu_5, \quad \varsigma_3 = \eta_4\mu_5 - \eta_5\mu_4.$$

This field separates the regular orbits.

Note that these theorems are valid for general ξ's. The special cases are considered in [10].

6.3.2   Flows on a Sphere

In this section we consider Euler system (6.4) on a two-dimensional unit sphere $M = S^2$ with the metric $g = \sin^2 y\, dx^2 + dy^2$ in the spherical coordinates.

The velocity field of the flow has the form $\mathbf{u} = u(t, x, y)\, \partial_x + v(t, x, y)\, \partial_y$, the pressure p, the density ρ, the temperature T, and the entropy s are the functions of time and space with the coordinates (t, x, y).

Here we consider the flow without any external force field, so $\mathbf{g} = 0$.

6.3.2.1 Symmetry Lie Algebra

As in the previous section, to describe the Lie algebra of symmetries, we consider the Lie algebra $\mathfrak{g}$ generated by the following vector fields on the manifold $J^0\pi$:

$$X_1 = \partial_t, \qquad X_2 = \partial_x,$$

$$X_3 = \frac{\cos x}{\tan y}\, \partial_x + \sin x\, \partial_y - \left(\frac{\sin x}{\tan y}\, u + \frac{\cos x}{\sin^2 y}\, v\right) \partial_u + u \cos x\, \partial_v,$$

$$X_4 = \frac{\sin x}{\tan y}\, \partial_x - \cos x\, \partial_y + \left(\frac{\cos x}{\tan y}\, u - \frac{\sin x}{\sin^2 y}\, v\right) \partial_u + u \sin x\, \partial_v,$$

$$X_5 = \partial_s, \qquad X_6 = \partial_p, \qquad X_7 = T\, \partial_T,$$

$$X_8 = t\, \partial_t - u\, \partial_u - v\, \partial_v + 2\rho\, \partial_\rho - s\, \partial_s,$$

$$X_9 = p\, \partial_p + \rho\, \partial_\rho - s\, \partial_s.$$

Consider the pure geometric and thermodynamic parts of this symmetry algebra.

The geometric part $\mathfrak{g}_m = \langle X_1, X_2, X_3, X_4 \rangle$ represents by the symmetries with respect to a group of sphere motions and time shifts, i.e. $\mathfrak{g}_m = \mathfrak{so}(3, \mathbb{R}) \oplus \mathbb{R}$, and $\mathfrak{g}_m = \ker \vartheta$.

To describe the thermodynamic part of the symmetry algebra, we denote by $\mathfrak{h}$ the Lie algebra generated by the vector fields

$$Y_1 = \partial_s, \qquad Y_2 = \partial_p, \qquad Y_3 = T\, \partial_T,$$

$$Y_4 = 2\rho\, \partial_\rho - s\, \partial_s, \qquad Y_5 = p\, \partial_p - \rho\, \partial_\rho.$$

This is a solvable Lie algebra with the following structure:

$$[Y_1, Y_4] = -Y_1, \qquad [Y_2, Y_5] = Y_2.$$

As above, let $\mathfrak{h}_t$ be the Lie subalgebra of algebra $\mathfrak{h}$ that preserves thermodynamic state (6.1).

Theorem 6.9 ([11]) *The Lie algebra $\mathfrak{g}_{sym}$ of point symmetries of the Euler system of differential equations on a two-dimensional unit sphere coincides with*

$$\vartheta^{-1}(\mathfrak{h}_t).$$

6.3.2.2 Symmetry Classification of States

In this section we classify the thermodynamic states or Lagrangian surfaces $\tilde{L}$ (compare with the previous section) depending on the dimension of the symmetry algebra $\mathfrak{h}_t \subset \mathfrak{h}$.

We consider one- and two-dimensional symmetry algebras only.

States with a One-Dimensional Symmetry Algebra

Let $\dim \mathfrak{h}_t = 1$ and let $Z = \sum_{i=1}^{5} \lambda_i Y_i$ be a basis vector in this algebra, then the differential 1-form $\iota_Z \Omega$ has the form

$$\iota_Z \Omega = \frac{2\lambda_4 - \lambda_5}{\rho} \, dp - \frac{\lambda_5 p + \lambda_2}{\rho^2} \, d\rho - \lambda_3 T \, ds + (\lambda_1 - \lambda_4 s) \, dT,$$

and the surface $\tilde{L}$ can be found from the following PDE system:

$$\begin{cases} (2\lambda_4 - \lambda_5)\rho \, \epsilon_{\rho\rho} + (\lambda_1 - \lambda_4 s)\epsilon_{\rho s} + (4\lambda_4 - 3\lambda_5) \, \epsilon_\rho - \dfrac{\lambda_2}{\rho^2} = 0, \\[2mm] (\lambda_1 - \lambda_4 s)\epsilon_{ss} + (2\lambda_4 - \lambda_5)\rho \, \epsilon_{\rho s} - \lambda_3 \, \epsilon_s = 0. \end{cases} \quad (6.9)$$

It is easy to check that the bracket of these two equations (see [8]) vanishes, and therefore the system is formally integrable and compatible.

Below we list solutions of this system under the assumption of parameters λ generality. A more detailed description may be found in [10, 11].

Solving the last system in case $\lambda_3 + \lambda_4 - 2\lambda_5 \neq 0$, we find the following expressions for the pressure and the temperature:

$$p = C_1 \rho^{\frac{\lambda_5}{2\lambda_4 - \lambda_5}} - \frac{\lambda_2}{\lambda_5}, \quad T = C_2(\lambda_1 - \lambda_4 s)^{-\frac{\lambda_3}{\lambda_4}},$$

where C_1, C_2 are constants.

The admissibility conditions (the negative definiteness of the form κ) have the form

$$\frac{\lambda_3}{\lambda_1 - \lambda_4 s} > 0, \quad \frac{\lambda_5 C_1}{2\lambda_4 - \lambda_5} > 0$$

for all $s \in (-\infty, s_0]$.

Theorem 6.10 *The thermodynamic states admitting a one-dimensional symmetry algebra have the form*

$$p = C_1 \rho^{\frac{\lambda_5}{2\lambda_4 - \lambda_5}} - \frac{\lambda_2}{\lambda_5}, \quad T = C_2(\lambda_1 - \lambda_4 s)^{-\frac{\lambda_3}{\lambda_4}},$$

where the constants defining the symmetry algebra satisfy inequalities

$$s_0 < \frac{\lambda_1}{\lambda_4}, \quad C_1 > 0, \quad \frac{\lambda_2}{\lambda_5} < 0, \quad \frac{\lambda_5}{2\lambda_4 - \lambda_5} > 0,$$

and besides they must meet one of the following conditions:

1. *if $\frac{\lambda_3}{\lambda_4}$ is irrational, then $\lambda_3 > 0$, $\lambda_4 > 0$, $C_2 > 0$;*
2. *if $\frac{\lambda_3}{\lambda_4}$ is rational, then $\frac{\lambda_3}{\lambda_4} > 0$ (i.e. $\frac{\lambda_3}{\lambda_4} = \frac{m}{k}$) and*

   a. *if k is even, then $\lambda_4 > 0$, $C_2 > 0$;*
   b. *if k is odd and m is even, then $C_2 > 0$;*
   c. *if k is odd and m is odd, then $C_2 \lambda_4 > 0$.*

States with a Two-Dimensional Symmetry Algebra

As in the plane case there are no thermodynamic states that admit a two-dimensional non-commutative symmetry algebra.

States with a Two-Dimensional Commutative Symmetry Algebra

Let now $\mathfrak{h}_t \subset \mathfrak{h}$ be a commutative two-dimensional Lie subalgebra, and let $A = \sum_{i=1}^{5} \alpha_i Y_i$, $B = \sum_{i=1}^{5} \beta_i Y_i$ be basis vectors in this algebra.
Then the condition $[A, B] = 0$ gives the following relations on α's and β's:

$$\alpha_1 \beta_4 - \alpha_4 \beta_1 = 0, \qquad \alpha_2 \beta_5 - \alpha_5 \beta_2 = 0. \tag{6.10}$$

Then, as above, restriction of the forms $\iota_A \Omega$ and $\iota_B \Omega$ on the state surface $\tilde{L}$ leads us to the four differential equations of the form (6.9), and the formal integrability condition for obtained system has the form

$$(\alpha_2 \beta_5 - \alpha_5 \beta_2)(5\beta_4 - 3\beta_5) = 0,$$

which is satisfied due to relations (6.10).
Solving this system for the general parameters α and β we get only the "nonphysical" solution of the form $\epsilon = C_1 \rho^{-1} + C_2$.
For the special case, for example,

$$\alpha_3 = \frac{\alpha_4 \beta_3}{\beta_4}, \quad \alpha_5 = \frac{\alpha_4 \beta_5}{\beta_4}$$

we get

$$p = C_1 \rho^{\frac{\beta_5}{2\beta_4 - \beta_5}} - \frac{\beta_2}{\beta_5}, \quad T = C_2 \left(s - \frac{\beta_1}{\beta_4} \right)^{-\frac{\beta_3}{\beta_4}}.$$

And the admissibility condition leads to the relations

$$\frac{\beta_3}{\beta_1 - \beta_4 s} > 0, \quad \frac{C_1 \beta_5}{2\beta_4 - \beta_5} > 0.$$

Theorem 6.11 *In the general case, there are no physically applicable thermodynamic states, which admit a two-dimensional commutative symmetry algebra.*

For the special case $\alpha_3 = \frac{\alpha_4 \beta_3}{\beta_4}$ *and* $\alpha_5 = \frac{\alpha_4 \beta_5}{\beta_4}$, *the thermodynamic states admitting a two-dimensional commutative symmetry algebra have the form*

$$p = C_1 \rho^{\frac{\beta_5}{2\beta_4 - \beta_5}} - \frac{\beta_2}{\beta_5}, \quad T = C_2 \left(s - \frac{\beta_1}{\beta_4} \right)^{-\frac{\beta_3}{\beta_4}},$$

where the constants defining the symmetry algebra satisfy inequalities

$$s_0 < \frac{\beta_1}{\beta_4}, \quad C_1 > 0, \quad \frac{\beta_2}{\beta_5} < 0, \quad \frac{\beta_5}{2\beta_4 - \beta_5} > 0, \quad \frac{\beta_3}{\beta_4} = \frac{m}{k} > 0,$$

i.e. $\frac{\beta_3}{\beta_4}$ *is rational positive number, and the following cases are possible:*

1. *if k is odd and m is even, then* $C_2 > 0$;
2. *if k is odd and m is odd, then* $C_2 \lambda_4 > 0$.

6.3.2.3 Differential Invariants

As in the previous section, we consider two group actions on the Euler equation $\mathcal{E}$, i.e. the prolonged action of the group generated by the action of the Lie algebra $\mathfrak{g}_m$ and the action generated by the prolongation of the action of the Lie algebra $\mathfrak{g}_{s\eta m}$. So we get two types of differential invariants—the kinematic and the Euler invariants.

The Field of Kinematic Invariants

First of all, the functions

$$\rho, \quad s, \quad g(\mathbf{u}, \mathbf{u})$$

(as well as p and T) generate all $\mathfrak{g}_m$-invariants of order zero.

Consider two vector fields $\mathbf{u}$ and $\tilde{\mathbf{u}}$ such that $g(\mathbf{u}, \tilde{\mathbf{u}}) = 0$ and $g(\mathbf{u}, \mathbf{u}) = g(\tilde{\mathbf{u}}, \tilde{\mathbf{u}})$. Writing the covariant differential $d_\nabla \mathbf{u}$ with respect to the vectors $\mathbf{u}$ and $\tilde{\mathbf{u}}$ as the sum of its symmetric and antisymmetric parts we obtain the 4 invariants of the first order:

$$J_1 = u_x + v_y + v \cot y, \qquad J_2 = u_y \sin y - \frac{v_x}{\sin y} + 2u \cos y,$$

$$J_3 = (u(u_x v - v_x u) + v(u_y v - v_y u)) \sin y + u \cos y (u^2 \sin^2 y + 2v^2),$$

$$J_4 = v(u_x v - v_x u) - u(u_y v - v_y u) \sin^2 y + v^3 \cot y.$$

$$(6.11)$$

The proof of the following theorem can be found in [11].

Theorem 6.12 ([11]) *The following derivations*

$$\nabla_1 = \frac{d}{dt}, \quad \nabla_2 = \frac{\rho_x}{\sin^2 y} \frac{d}{dx} + \rho_y \frac{d}{dy}, \quad \nabla_3 = \frac{s_x}{\sin^2 y} \frac{d}{dx} + s_y \frac{d}{dy}$$

are $\mathfrak{g}_m$*-invariant. They are linearly independent if*

$$\rho_x s_y - \rho_y s_x \neq 0.$$

It is easy to check that the codimension of regular $\mathfrak{g}_m$-orbits is equal to 12. The Rosenlicht theorem [17] gives us the following result.

Theorem 6.13 ([11]) *The field of the first order kinematic invariants is generated by the invariants* ρ, s, $g(\mathbf{u}, \mathbf{u})$ *of order zero and by the invariants* (6.11) *and*

$$\nabla_1 \rho, \quad \nabla_1 s, \quad \nabla_2 \rho, \quad \nabla_2 s, \quad \nabla_3 s \qquad (6.12)$$

of order one. These invariants separate regular $\mathfrak{g}_m$*-orbits.*

The bundle $\pi_{2,1} : \mathcal{E}_2 \to \mathcal{E}_1$ has rank 14, and by applying the derivations $\nabla_1, \nabla_2, \nabla_3$ to the kinematic invariants (6.11) and (6.12) we get 27 kinematic invariants. Straightforward computations show that among these invariants 14 are always independent (see https://d-omega.org).

Therefore, starting with the order $k = 1$ dimensions of regular orbits are equal to $\dim \mathfrak{g}_m = 4$.

The Hilbert function (the number of independent invariants) of the $\mathfrak{g}_m$-invariants field has form $H(k) = 5k + 4$ for $k \geqslant 1$ and $H(0) = 3$, and the corresponding Poincaré function is equal to

$$P(z) = \frac{3 + 3z - z^3}{(1 - z)^2}.$$

Summarizing, we get the following result.

Theorem 6.14 ([11]) *The field of the kinematic invariants is generated by the invariants ρ, s, $g(\mathbf{u}, \mathbf{u})$ of order zero, by the invariants (6.11) and (6.12) of order one, and by the invariant derivations ∇_1, ∇_2, ∇_3. This field separates regular orbits.*

The Field of Euler Invariants

Let us consider the case when the equations of thermodynamic state $\tilde{L}$ admit a one-dimensional symmetry algebra generated by the vector field

$$A = \xi_1 X_5 + \xi_2 X_6 + \xi_3 X_7 + \xi_4 X_8 + \xi_5 X_9.$$

Using a similar computation as in the plane case we get the following result.

Theorem 6.15 ([11]) *The field of the Euler differential invariants on a sphere for thermodynamic states admitting a one-dimensional symmetry algebra is generated by the differential invariants*

$$J_1 \rho \left(s - \frac{\xi_1}{\xi_4 + \xi_5} \right), \quad J_1 \rho^{\frac{\xi_4}{2\xi_4 + \xi_5}}, \quad \frac{g(\mathbf{u}, \mathbf{u})}{J_1^2},$$

$$\frac{J_2}{J_1}, \quad \frac{J_3}{J_1^3}, \quad \frac{J_4}{J_1^3}, \quad \frac{\nabla_1 \rho}{J_1 \rho}, \quad \frac{\nabla_2 \rho}{\rho^2}, \quad \rho \nabla_1 s, \quad J_1 \nabla_2 s, \quad J_1^2 \rho^2 \nabla_3 s$$

of the first order and by the invariant derivations

$$\rho^{\frac{\xi_4}{2\xi_4 + \xi_5}} \nabla_1, \quad \rho^{-1} \nabla_2, \quad \rho^{\frac{\xi_4 + \xi_5}{2\xi_4 + \xi_5}} \nabla_3.$$

This field separates regular orbits.

The last formulas are valid for general ξ's. All details and the special cases are considered in [11].

Now let the thermodynamic state admit a commutative two-dimensional symmetry algebra generated by the vector fields $A = \sum_{i=1}^{5} \mu_i X_{i+4}$, $B = \sum_{i=1}^{5} \eta_i X_{i+4}$ such that μ's and η's satisfy relations

$$\begin{cases} \eta_1 \mu_4 - \eta_4 \mu_1 + \eta_1 \mu_5 - \eta_5 \mu_1 = 0, \\ \eta_2 \mu_5 - \eta_5 \mu_2 = 0. \end{cases}$$

Theorem 6.16 ([11]) *The field of Euler differential invariants for the thermodynamic states admitting a commutative two-dimensional symmetry algebra is generated by differential invariants*

$$J_1\rho((\mu_4+\mu_5)s-\mu_1),\quad \frac{g(\mathbf{u},\mathbf{u})}{J_1^2},\quad \frac{J_2}{J_1},\quad \frac{J_3}{J_1^3},\quad \frac{J_4}{J_1^3},$$

$$\frac{\nabla_1\rho}{J_1\rho},\quad \rho\nabla_1 s,\quad \frac{\nabla_2\rho}{\rho^2},\quad J_1\nabla_2 s,\quad J_1^2\rho^2\nabla_3 s$$

of the first order and by the invariant derivations

$$\rho((\mu_4+\mu_5)s-\mu_1)\nabla_1,\quad \rho^{-1}\nabla_2,\quad ((\mu_4+\mu_5)s-\mu_1)^{-1}\nabla_3.$$

This field separates regular orbits.

6.3.3 Flows on a Spherical Layer

Consider Euler system (6.4) on a spherical layer $M = S^2 \times \mathbb{R}$ with the coordinates (x, y, z), where (x, y) are the stereographic coordinates on the sphere, and the metric

$$g = \frac{4}{(x^2 + y^2 + 1)^2}(dx^2 + dy^2) + dz^2.$$

The velocity field of the flow has the form $\mathbf{u} = u(t, x, y, z)\,\partial_x + v(t, x, y, z)\,\partial_y + w(t, x, y, z)\,\partial_z$, the pressure p, the density ρ, the temperature T, and the entropy s are the functions of time and space with the coordinates (t, x, y, z).

The vector of gravitational acceleration is of the form $\mathbf{g} = (0, 0, g)$.

6.3.3.1 Symmetry Lie Algebra

Consider the Lie algebra $\mathfrak{g}$ generated by the following vector fields on the manifold $J^0\pi$:

$$X_1 = \partial_t, \qquad X_3 = t\,\partial_z + \partial_w,$$

$$X_2 = \partial_z, \qquad X_4 = y\,\partial_x - x\,\partial_y + v\,\partial_u - u\,\partial_v,$$

$$X_5 = xy\,\partial_x - \frac{1}{2}(x^2 - y^2 - 1)\partial_y + (xv + yu)\,\partial_u - (xu - yv)\partial_v,$$

$$X_6 = -\frac{1}{2}(x^2 - y^2 + 1)\partial_x + xy\,\partial_y + (xu - yv)\partial_u + (xv + yu)\partial_v,$$

$$X_7 = \partial_s, \qquad X_{10} = t\,\partial_t + gt^2\,\partial_z - u\,\partial_u - v\,\partial_v + (2gt - w)\partial_w + 2\rho\,\partial_\rho - s\,\partial_s,$$

$$X_8 = \partial_p, \qquad X_{11} = p\,\partial_p + \rho\,\partial_\rho - s\,\partial_s,$$

$$X_9 = T\,\partial_T.$$

The pure geometric part $\mathfrak{g}_m$ generated by the vector fields $X_1, X_2, \ldots, X_6$. Transformations corresponding to the elements of the Lie group generated by the algebra $\mathfrak{g}_m$ are compositions of sphere motions, Galilean transformations, and shifts along the z direction, time shifts.

To describe thermodynamic part of the symmetry algebra, we consider the Lie algebra $\mathfrak{h}$ generated by the vector fields

$$Y_1 = \partial_s, \qquad Y_2 = \partial_p, \qquad Y_3 = T\,\partial_T,$$
$$Y_4 = 2\rho\,\partial_\rho - s\,\partial_s, \qquad Y_5 = p\,\partial_p - \rho\,\partial_\rho.$$

This is a solvable Lie algebra with the following structure:

$$[Y_1, Y_4] = -Y_1, \qquad [Y_2, Y_5] = Y_2.$$

Let also $\mathfrak{h}_t$ be the Lie subalgebra of algebra $\mathfrak{h}$ that preserves thermodynamic state (6.1). Then the following result is valid.

Theorem 6.17 ([12]) *The Lie algebra $\mathfrak{g}_{s\eta m}$ of point symmetries of the Euler system of differential equations on a spherical layer coincides with*

$$\vartheta^{-1}(\mathfrak{h}_t).$$

6.3.3.2 Symmetry Classification of States

The Lie algebra generated by the vector fields $Y_1, \ldots, Y_5$ coincides with the Lie algebra of the thermodynamic symmetries of the Euler system on a sphere.

Thus the classification of the thermodynamic states or Lagrangian surfaces $\tilde{L}$ depending on the dimension of the symmetry algebra $\mathfrak{h}_t \subset \mathfrak{h}$ is the same as the classification presented in the previous section.

6.3.3.3 Differential Invariants

The Field of Kinematic Invariants

First of all, the functions ρ, s, $g(\mathbf{u}, \mathbf{u}) - w^2$ (as well as p and T) generate all $\mathfrak{g}_m$-invariants of order zero.

The proofs of the following theorems can be found in [12].

Theorem 6.18 ([12]) *The following derivations*

$$\nabla_1 = \frac{\mathrm{d}}{\mathrm{d}z}, \qquad \nabla_2 = \frac{\mathrm{d}}{\mathrm{d}t} + w\frac{\mathrm{d}}{\mathrm{d}z}, \qquad \nabla_3 = u\frac{\mathrm{d}}{\mathrm{d}x} + v\frac{\mathrm{d}}{\mathrm{d}y}, \qquad \nabla_4 = v\frac{\mathrm{d}}{\mathrm{d}x} - u\frac{\mathrm{d}}{\mathrm{d}y}$$

are $\mathfrak{g}_m$-*invariant. They are linearly independent if*

$$u^2 + v^2 \neq 0.$$

Theorem 6.19 ([12]) *The field of the first order kinematic invariants is generated by the invariants* ρ, s, $g(\mathbf{u}, \mathbf{u}) - w^2$ *of order zero and by the invariants*

$$\nabla_1 \rho, \quad \nabla_2 \rho, \quad \nabla_3 \rho, \quad \nabla_4 \rho, \quad \nabla_1 s, \quad \nabla_2 s, \quad \nabla_3 s, \quad \nabla_4 s,$$

$$\nabla_1(g(\mathbf{u}, \mathbf{u}) - w^2), \quad \nabla_3(g(\mathbf{u}, \mathbf{u}) - w^2), \quad \nabla_4(g(\mathbf{u}, \mathbf{u}) - w^2), \qquad (6.13)$$

$$\nabla_1 w, \quad \nabla_3 w, \quad \nabla_4 w, \quad J_1 = u_z w_x + v_z w_y, \quad J_2 = \frac{u_t v_z - u_z v_t}{u_z^2 + v_z^2}$$

of order one. These invariants separate regular $\mathfrak{g}_m$-*orbits.*

The bundle $\pi_{2,1} : \mathcal{E}_2 \to \mathcal{E}_1$ has rank 33, and by applying the derivations ∇_i, $i = 1, \ldots, 4$ to the first order kinematic invariants (6.13) we get 64 kinematic invariants. Straightforward computations show that among these invariants 33 are always independent.

Therefore, starting with the order $k = 1$ dimensions of the regular orbits are equal to $\dim \mathfrak{g}_m = 6$.

Moreover, the number of independent invariants (the Hilbert function) is equal to $H(k) = 3k^2 + 8k + 5$ for $k \geqslant 1$ and $H(0) = 3$.

The corresponding Poincaré function has the form

$$P(z) = \frac{3 + 7z - 6z^2 + 2z^3}{(1 - z)^3}.$$

Theorem 6.20 ([12]) *The field of the kinematic invariants is generated by the invariants* ρ, s, $g(\mathbf{u}, \mathbf{u}) - w^2$ *of order zero, by the invariants* (6.13) *of order one, and by the invariant derivations* ∇_i, $i = 1, \ldots, 4$. *This field separates regular orbits.*

The Field of Euler Invariants

At first we consider the case when the thermodynamic state $\tilde{L}$ admits a one-dimensional symmetry algebra generated by the vector field

$$A = \xi_1 X_7 + \xi_2 X_8 + \xi_3 X_9 + \xi_4 X_{10} + \xi_5 X_{11}.$$

Then for general values of the parameters ξ's we have the following result. The special cases are considered in [12].

Theorem 6.21 ([12]) *The field of the Euler differential invariants for thermodynamic states admitting a one-dimensional symmetry algebra is generated by the differential invariants*

$$w_z \rho \left(s - \frac{\xi_1}{\xi_4 + \xi_5} \right), \quad w_z^{-2} \left(g(\mathbf{u}, \mathbf{u}) - w^2 \right),$$

$$\frac{\nabla_1 \rho}{\rho}, \quad \frac{\nabla_2 \rho}{w_z \rho}, \quad \frac{\nabla_3 \rho}{w_z \rho}, \quad \frac{\nabla_4 \rho}{w_z \rho},$$

$$w_z \rho \nabla_1 s, \quad \rho \nabla_2 s, \quad \rho \nabla_3 s, \quad \rho \nabla_4 s,$$

$$w_z^{-2} \nabla_1 \left(g(\mathbf{u}, \mathbf{u}) - w^2 \right), \quad w_z^{-3} \nabla_3 \left(g(\mathbf{u}, \mathbf{u}) - w^2 \right), \quad w_z^{-3} \nabla_4 \left(g(\mathbf{u}, \mathbf{u}) - w^2 \right),$$

$$w_z \rho^{\frac{\xi_4}{2\xi_4 + \xi_5}}, \quad w_z^{-2} \nabla_3 w, \quad w_z^{-2} \nabla_4 w, \quad w_z^{-2} J_1, \quad w_z^{-1} J_2,$$

of the first order and by the invariant derivatives

$$\nabla_1, \quad w_z^{-1} \nabla_2, \quad w_z^{-1} \nabla_3, \quad w_z^{-1} \nabla_4.$$

This field separates regular orbits.

Now, let the thermodynamic state admit a commutative two-dimensional symmetry algebra generated by the vector fields $A = \sum_{i=1}^{5} \mu_i X_{i+6}$, $B = \sum_{i=1}^{5} \eta_i X_{i+6}$, then μ's and η's satisfy relations

$$\begin{cases} \eta_1 \mu_4 - \eta_4 \mu_1 + \eta_1 \mu_5 - \eta_5 \mu_1 = 0, \\ \eta_2 \mu_5 - \eta_5 \mu_2 = 0. \end{cases}$$

Theorem 6.22 ([12]) *The field of Euler differential invariants for thermodynamic states admitting a commutative two-dimensional symmetry algebra is generated by differential invariants*

$$w_z^{-2} \left(g(\mathbf{u}, \mathbf{u}) - w^2 \right),$$

$$\frac{\nabla_1 \rho}{\rho}, \quad \frac{\nabla_2 \rho}{w_z \rho}, \quad \frac{\nabla_3 \rho}{w_z \rho}, \quad \frac{\nabla_4 \rho}{w_z \rho},$$

$$w_z \rho \nabla_1 s, \quad \rho \nabla_2 s, \quad \rho \nabla_3 s, \quad \rho \nabla_4 s,$$

$$w_z^{-2} \nabla_1 \left(g(\mathbf{u}, \mathbf{u}) - w^2 \right), \quad w_z^{-3} \nabla_3 \left(g(\mathbf{u}, \mathbf{u}) - w^2 \right), \quad w_z^{-3} \nabla_4 \left(g(\mathbf{u}, \mathbf{u}) - w^2 \right),$$

$$w_z \rho ((\mu_4 + \mu_5) s - \mu_1), \quad w_z^{-2} \nabla_3 w, \quad w_z^{-2} \nabla_4 w, \quad w_z^{-2} J_1, \quad w_z^{-1} J_2$$

of the first order and by the invariant derivatives

$$\nabla_1, \quad w_z^{-1} \nabla_2, \quad w_z^{-1} \nabla_3, \quad w_z^{-1} \nabla_4.$$

This field separates regular orbits.

6.4 Compressible Viscid Fluids or Gases

In this section we study differential invariants of compressible viscid fluids or gases.

The system of differential equations (the Navier–Stokes system) describing flows on an oriented Riemannian manifold (M, g) consists of the following equations (see [1] for details):

$$
\begin{cases}
\rho(\mathbf{u}_t + \nabla_{\mathbf{u}}\mathbf{u}) - \operatorname{div}\sigma - \mathbf{g}\rho = 0, \\
\dfrac{\partial(\rho\,\Omega_g)}{\partial t} + \mathcal{L}_{\mathbf{u}}\left(\rho\,\Omega_g\right) = 0, \\
\rho T\,(s_t + \nabla_{\mathbf{u}}s) - \Phi + k(\Delta_g T) = 0.
\end{cases}
\tag{6.14}
$$

Here the divergence operator $\operatorname{div} : S^2 T^* M \to TM$ is given by

$$
(\operatorname{div}\sigma)_l = (d_\nabla\sigma)_{ijk}g^{jk}g^{il},
$$

where d_∇ is the covariant differential.

The fluid under consideration is assumed to be Newtonian and isotropic. Therefore, the fluid stress tensor σ is symmetric, and it depends on the rate of deformation tensor $D = \frac{1}{2}\mathcal{L}_{\mathbf{u}}(g)$ linearly. These two conditions give the following form of the stress tensor: $\sigma = -pg + o'$, where the viscous stress tensor σ' is given by

$$
\sigma' = 2\eta\left(D - \frac{\langle D, g\rangle_g}{\langle g, g\rangle_g}g\right) + \zeta\langle D, g\rangle_g g.
$$

The quantity $\Phi = \langle \sigma', D\rangle_g$ represents the rate of dissipation of mechanical energy [1].

The first equation of system (6.14) is the Navier–Stokes equation, the second one is the continuity equation, and the third one is the general equation of heat transfer.

In this section we consider the following examples of manifold M: a plane, a three-dimensional space, a sphere, and a spherical layer.

Note that in all these cases the number of unknown functions is greater than the number of system equations by 2, i.e. the system (6.14) is incomplete. As above we get two additional equations using the thermodynamics of the medium.

Thus, by the Navier–Stokes system of differential equations we mean the system (6.14) extended by two equations of state (6.1), where functions f and g satisfy the additional relation (6.2) and the form κ is negative definite.

Geometrically, we represent this system in the following way. Consider the bundle

$$
\pi : \mathbb{R} \times TM \times \mathbb{R}^4 \longrightarrow \mathbb{R} \times M
$$

of rank (dim $M + 4$).

Then the Navier–Stokes system is a system of differential equations on sections of the bundle π.

Note that system (6.1) defines the zeroth order system $\mathcal{E}_0 \subset J^0\pi$.

Denote by $\mathcal{E}_1 \subset J^1\pi$ the system of order $\leqslant 1$ obtained by the first prolongation of the system $\mathcal{E}_0$ and by the continuity equation of system (6.14).

Let also $\mathcal{E}_2 \subset J^2\pi$ be the system of differential equations of order $\leqslant 2$ obtained by the first prolongation of the system $\mathcal{E}_1$ and all equations of system (6.14).

For the case $k \geqslant 3$, we define $\mathcal{E}_k \subset J^k\pi$ to be the $(k-2)$-th prolongation of the system $\mathcal{E}_2$.

Note that the system $\mathcal{E}_\infty = \varprojlim \mathcal{E}_k$ is a formally integrable system of differential equations, which we also call the Navier–Stokes system.

6.4.1   2D-Flows

Consider Navier–Stokes system (6.14) on a plane $M = \mathbb{R}^2$ equipped with the coordinates (x, y) and the standard flat metric $g = dx^2 + dy^2$.

The velocity field of the flow has the form $\mathbf{u} = u(t, x, y)\,\partial_x + v(t, x, y)\,\partial_y$, the pressure p, the density ρ, the temperature T, and the entropy s are the functions of time and space with the coordinates (t, x, y).

Here we also consider the flow without any external force field, so $\mathbf{g} = 0$.

6.4.1.1   Symmetry Lie Algebra

To describe the Lie algebra of symmetries of the Navier–Stokes system we consider a Lie algebra $\mathfrak{g}$ generated by the following vector fields on space $J^0\pi$:

$$X_1 = \partial_x, \qquad\qquad\qquad X_4 = t\,\partial_x + \partial_u,$$
$$X_2 = \partial_y, \qquad\qquad\qquad X_5 = t\,\partial_y + \partial_v,$$
$$X_3 = y\,\partial_x - x\,\partial_y + v\,\partial_u - u\,\partial_v, \qquad X_6 = \partial_t,$$
$$X_7 = \partial_s, \qquad X_8 = \partial_p,$$
$$X_9 = x\,\partial_x + y\,\partial_y + u\,\partial_u + v\,\partial_v - 2\rho\,\partial_\rho + 2T\,\partial_T,$$
$$X_{10} = t\,\partial_t - u\,\partial_u - v\,\partial_v + \rho\,\partial_\rho - p\,\partial_p - 2T\,\partial_T.$$

In general the symmetry algebra of system (6.14) consists of pure geometric and thermodynamic parts.

The geometric part is represented by the algebra $\mathfrak{g}_m = \langle X_1, X_2, \ldots, X_6 \rangle$ with respect to the group of motions, Galilean transformations, and time shifts.

Moreover, the kernel of homomorphism ϑ (6.5) is an ideal $\mathfrak{g}_m$ in the Lie algebra $\mathfrak{g}$.

The thermodynamic part strongly depends on the symmetries of the thermodynamic state. In order to describe it, denote by $\mathfrak{h}$ a Lie algebra generated by the vector fields

$$Y_1 = \partial_s, \quad Y_2 = \partial_p, \quad Y_3 = \rho\,\partial_\rho - T\,\partial_T, \quad Y_4 = p\,\partial_p + T\,\partial_T.$$

Let also $\mathfrak{h}_t$ be a Lie subalgebra of the algebra $\mathfrak{h}$ which preserves the thermodynamic state (6.1).

Theorem 6.23 ([13]) *A Lie algebra $\mathfrak{g}_{s\eta m}$ of symmetries of the Navier–Stokes system of differential equations on a plane coincides with*

$$\vartheta^{-1}(\mathfrak{h}_t).$$

Note that, usually, the equations of state are neglected and the vector fields like $f(t)\,\partial_p$, where f is an arbitrary function, considered as symmetries of the Navier–Stokes system.

For the general equation of state $\mathfrak{h}_t = 0$ and the symmetry algebra coincides with the algebra $\mathfrak{g}_m$.

6.4.1.2 Symmetry Classification of States

In this section we classify thermodynamic states or Lagrangian surfaces $\tilde{L}$ depending on the dimension of the symmetry algebra $\mathfrak{h}_t \subset \mathfrak{h}$.

We consider one- and two-dimensional symmetry algebras only. One can easily check that there are no physically valuable thermodynamic states with three or more dimensional symmetry algebras.

States with a One-Dimensional Symmetry Algebra

Let $\dim \mathfrak{h}_t = 1$ and let $Z = \sum_{i=1}^{4} \lambda_i Y_i$ be a basis vector in this algebra, then the differential 1-form $\iota_Z \Omega$ has the form

$$\iota_Z \Omega = -\frac{\lambda_3}{\rho}\,dp + \frac{\lambda_4 p + \lambda_2}{\rho^2}\,d\rho + (\lambda_3 - \lambda_4)T\,ds - \lambda_1 dT,$$

and, in terms of specific energy $\epsilon(\rho, s)$, the Lagrangian surface $\tilde{L}$ can be found as a solution of the following PDE system:

$$\begin{cases} \lambda_1 \epsilon_{ss} + \lambda_3 \rho \epsilon_{s\rho} + (\lambda_3 - \lambda_4)\epsilon_s = 0, \\[2mm] \lambda_3 \rho \epsilon_{\rho\rho} + \lambda_1 \epsilon_{s\rho} + (2\lambda_3 - \lambda_4)\epsilon_\rho - \dfrac{\lambda_2}{\rho^2} = 0. \end{cases}$$

It is easy to check that the bracket of these two equations (see [8]) vanishes and therefore the system is formally integrable and compatible.

Solving this system for general values of parameters λ, all special cases are considered in [13], we get expressions for the presser and the temperature

$$T = \rho^{\frac{\lambda_4}{\lambda_3}-1} F', \quad p = \rho^{\frac{\lambda_4}{\lambda_3}} \left(\left(\frac{\lambda_4}{\lambda_3} - 1 \right) F - \frac{\lambda_1}{\lambda_3} F' \right) - \frac{\lambda_2}{\lambda_4}, \quad F = F\left(s - \frac{\lambda_1}{\lambda_3} \ln \rho \right),$$

where F is a smooth function.

Negative definiteness of the quadratic form κ gives the following relations on the function F and the parameters λ:

$$\lambda_1^2 \rho^{\frac{\lambda_4}{\lambda_3}} F'' + \lambda_1(\lambda_3 - \lambda_4)\rho T + \lambda_3(\lambda_4 p + \lambda_2) > 0,$$

$$\rho^{\frac{\lambda_4 - 2\lambda_3}{\lambda_3}} F'' (\lambda_1(\lambda_3 - \lambda_4)\rho T - \lambda_3(\lambda_4 p + \lambda_2)) + T^2(\lambda_3 - \lambda_4)^2 < 0.$$

Theorem 6.24 *The thermodynamic states admitting a one-dimensional symmetry algebra have the form*

$$T = \rho^{\frac{\lambda_4}{\lambda_3}-1} F', \quad p = \rho^{\frac{\lambda_4}{\lambda_3}} \left(\left(\frac{\lambda_4}{\lambda_3} - 1 \right) F - \frac{\lambda_1}{\lambda_3} F' \right) - \frac{\lambda_2}{\lambda_4}, \quad F = F\left(s - \frac{\lambda_1}{\lambda_3} \ln \rho \right),$$

where F is a smooth function, F' is positive, and

$$\lambda_1^2 F'' + \lambda_1(\lambda_3 - 2\lambda_4) F' + \lambda_4(\lambda_4 - \lambda_3) F > 0,$$

$$F''(\lambda_4(\lambda_4 - \lambda_3) F - \lambda_1 \lambda_3 F') - (F')^2(\lambda_4 - \lambda_3)^2 > 0.$$

States with a Two-Dimensional Non-commutative Symmetry Algebra

Let $\mathfrak{h}_t \subset \mathfrak{h}$ be a non-commutative two-dimensional Lie subalgebra. It is easy to check that two vectors of the form $A = Y_2$, $B = \alpha Y_1 + \beta Y_3 + Y_4$ are the basis vectors in the non-commutative algebra $\mathfrak{h}_t$.

Then, as above, the restrictions of forms $\iota_A \Omega$ and $\iota_B \Omega$ on the state surface $\tilde{L}$ lead us to the solution $\rho = const$.

Since we consider thermodynamic states such that the variables ρ and s are local coordinates then we do not consider the case of the non-commutative subalgebra.

States with a Two-Dimensional Commutative Symmetry Algebra

Let now $\mathfrak{h}_t \subset \mathfrak{h}$ be a commutative two-dimensional Lie subalgebra, and let $A = \sum_{i=1}^{4} \alpha_i Y_i$, $B = \sum_{i=1}^{4} \beta_i Y_i$ be the basis vectors in the algebra $\mathfrak{h}_t$.

Then condition $[A, B] = 0$ gives the following relations on α's and β's:

$$\alpha_2 \beta_4 - \alpha_4 \beta_2 = 0. \tag{6.15}$$

Then the restrictions of forms $\iota_A \Omega$ and $\iota_B \Omega$ on the state surface $\tilde{L}$ lead us to the following system of differential equations:

$$\begin{cases} \alpha_1 \epsilon_{ss} + \alpha_3 \rho \epsilon_{s\rho} + (\alpha_3 - \alpha_4)\epsilon_s = 0, \\[2mm] \alpha_3 \rho \epsilon_{\rho\rho} + \alpha_1 \epsilon_{s\rho} + (2\alpha_3 - \alpha_4)\epsilon_\rho - \dfrac{\alpha_2}{\rho^2} = 0, \\[2mm] \beta_1 \epsilon_{ss} + \beta_3 \rho \epsilon_{s\rho} + (\beta_3 - \beta_4)\epsilon_s = 0, \\[2mm] \beta_3 \rho \epsilon_{\rho\rho} + \beta_1 \epsilon_{s\rho} + (2\beta_3 - \beta_4)\epsilon_\rho - \dfrac{\beta_2}{\rho^2} = 0. \end{cases}$$

The formal integrability condition for this system has the form

$$(5\beta_3 - \beta_4)(\alpha_2 \beta_4 - \beta_2 \alpha_4) = 0,$$

which is satisfied due to relations (6.15).

Solving this PDE system we get the following expressions for the pressure and the temperature:

$$p = C(\beta - 1)e^{\alpha s} \rho^\beta - \frac{\beta_2}{\beta_4}, \quad T = C\alpha e^{\alpha s} \rho^{\beta-1}, \tag{6.16}$$

where

$$\alpha = \frac{\alpha_4 \beta_3 - \alpha_3 \beta_4}{\alpha_1 \beta_3 - \alpha_3 \beta_1}, \quad \beta = \frac{\alpha_1 \beta_4 - \beta_1 \alpha_4}{\alpha_1 \beta_3 - \alpha_3 \beta_1},$$

and the admissibility conditions have the form $\alpha > 0$, $\beta > 1$

Theorem 6.25 *The thermodynamic states admitting a two-dimensional commuta ilve symmetry algebra have the form*

$$p = C(\beta - 1)e^{\alpha s} \rho^\beta - \frac{\beta_2}{\beta_4}, \quad T = C\alpha e^{\alpha s} \rho^{\beta-1},$$

where

$$\alpha = \frac{\alpha_4\beta_3 - \alpha_3\beta_4}{\alpha_1\beta_3 - \alpha_3\beta_1} > 0, \quad \beta = \frac{\alpha_1\beta_4 - \beta_1\alpha_4}{\alpha_1\beta_3 - \alpha_3\beta_1} > 1, \quad C > 0, \quad \frac{\beta_2}{\beta_4} < 0.$$

Observe that, the expressions for the temperature and the pressure for an ideal gas

$$T = \frac{1}{\gamma}\rho^k e^{\frac{s}{\gamma}}, \quad p = k\rho^{k+1} e^{\frac{s}{\gamma}},$$

where k and γ are constant depending on a gas, can be obtained from the equations (6.16) by choosing appropriate values of the constants.

6.4.1.3 Differential Invariants

As in the case of compressible inviscid fluids or gases (the Euler system), we consider two group actions on the Navier–Stokes equation $\mathcal{E}$.

The first one is the prolonged action of the group generated by the action of Lie algebra $\mathfrak{g}_m$. The differential invariants with respect to this action are called *kinematic differential invariants*.

The second one is the prolonged action of the group generated by the action of Lie algebra $\mathfrak{g}_{s\eta m}$. The corresponding differential invariants are called *Navier–Stokes invariants*.

Also we say that a point $x_k \in \mathcal{E}_k$ and the corresponding orbit $\mathcal{O}(x_k)$ ($\mathfrak{g}_m$ or $\mathfrak{g}_{s\eta m}$-orbit) are *regular*, if there are exactly $m = \text{codim}\,\mathcal{O}(x_k)$ independent invariants (kinematic or Navier–Stokes) in a neighborhood of this orbit.

Thus, the corresponding point on the quotient space $\mathcal{E}_k/\mathfrak{g}_m$ or $\mathcal{E}_k/\mathfrak{g}_{s\eta m}$ is smooth, and these independent invariants (kinematic or Navier–Stokes) can serve as local coordinates in a neighborhood of this point.

Otherwise, we say that the point and the corresponding orbit are *singular*.

It is worth to note that the Navier–Stokes system together with the symmetry algebras $\mathfrak{g}_m$ or $\mathfrak{g}_{s\eta m}$ satisfies the conditions of the Lie–Tresse theorem (see [9]), and therefore the above differential invariants separate regular $\mathfrak{g}_m$ or $\mathfrak{g}_{s\eta m}$ orbits on the Navier–Stokes system $\mathcal{E}$.

The Field of Kinematic Invariants

First of all observe that the density ρ and the entropy s (as well as the pressure p and the temperature T) on the equation $\mathcal{E}_0$ are $\mathfrak{g}_m$-invariants.

Moreover, the following functions are the kinematic invariants of the first order (see [13]):

$$J_1 = u_x + v_y, \qquad J_5 = \rho_x s_y - \rho_y s_x,$$

$$J_2 = u_y - v_x, \qquad J_6 = s_t + s_x u + s_y v,$$

$$J_3 = \rho_x^2 + \rho_y^2, \qquad J_7 = \rho_x(\rho_x u_x + \rho_y u_y) + \rho_y(\rho_x v_x + \rho_y v_y),$$

$$J_4 = s_x^2 + s_y^2, \qquad J_8 = s_x(\rho_x u_x + \rho_y u_y) + s_y(\rho_x v_x + \rho_y v_y),$$

$$J_9 = s_x(u_t + u u_x + v u_y) + s_y(v_t + u v_x + v v_y),$$

$$J_{10} = \rho_x(u_t + u u_x + v u_y) + \rho_y(v_t + u v_x + v v_y).$$

Proposition 6.2 *The singular points belong to the union of two sets:*

$$\Upsilon_1 = \{ u_x - v_y = 0, \ u_y + v_x = 0, \ u_t = v_t = \rho_x = \rho_y = s_x = s_y = 0 \},$$

$$\Upsilon_2 = \{ J_3 J_5^2 (J_3 J_4 - J_5^2) = 0 \}.$$

The set Υ_1 contains singular points that have five-dimensional orbits. The set Υ_2 contains points where differential invariants $J_1, J_4, \ldots, J_{10}$ are dependent.

It is easy to check that codimension of regular $\mathfrak{g}_m$-orbits is equal to 12. The proofs of the following theorems can be found in [13].

Theorem 6.26 ([13]) *The field of the first order kinematic invariants is generated by invariants $\rho, s, J_1, \ldots, J_{10}$. These invariants separate regular $\mathfrak{g}_m$-orbits.*

Theorem 6.27 ([13]) *The following derivations*

$$\nabla_1 = \frac{d}{dt} + u\frac{d}{dx} + v\frac{d}{dy}, \quad \nabla_2 = \rho_x\frac{d}{dx} + \rho_y\frac{d}{dy}, \quad \nabla_3 = s_x\frac{d}{dx} + s_y\frac{d}{dy}$$

are $\mathfrak{g}_m$-invariant. They are linear independent if

$$\rho_x s_y - \rho_y s_x \neq 0.$$

The bundle $\pi_{2,1} : \mathcal{E}_2 \rightarrow \mathcal{E}_1$ has rank 18 and by applying derivations $\nabla_1, \nabla_2, \nabla_3$ to the kinematic invariants $J_1, J_2, \ldots, J_{10}$ we get 30 kinematic invariants. Straightforward computations show that among these invariants 18 are always independent (see https://d-omega.org).

Therefore, beginning with order $k - 1$ dimensions of regular orbits are equal to $\dim \mathfrak{g}_m - 6$.

Moreover, the number of independent invariants of pure order k (the Hilbert function) is equal to $H(k) = 7k + 4$ for $k \geqslant 2$, and $H(0) = 2$, $H(1) = 10$.

The corresponding Poincaré function is equal to

$$P(z) = \frac{2 + 6z - z^3}{(1 - z)^2}.$$

Theorem 6.28 ([13]) *The field of kinematic invariants is generated by the invariants ρ, s of order zero, invariants $J_1, J_2, \ldots, J_{10}$ of order one, and by the invariant derivations $\nabla_1, \nabla_2, \nabla_3$. This field separates regular orbits.*

The Field of Navier–Stokes Invariants

Here we consider the case when the thermodynamic state admits a one-dimensional symmetry algebra generated by the vector field

$$A = \xi_1 X_7 + \xi_2 X_8 + \xi_3 X_9 + \xi_4 X_{10}.$$

Note that, the $\mathfrak{g}_m$ invariant derivations $\nabla_1, \nabla_2, \nabla_3$ do not commute with the thermodynamic symmetry A. Moreover, the action of the thermodynamic vector field A on the field of kinematic invariants is given by the following derivation:

$$(\xi_4 - 2\xi_3)\rho\partial_\rho + \xi_1\partial_s - \xi_4 J_1\partial_{J_1} - \xi_4 J_2\partial_{J_2} - 2J_3(3\xi_3 - \xi_4)\partial_{J_3} -$$

$$-2\xi_3 J_4\partial_{J_4} + J_5(\xi_4 - 4\xi_3)\partial_{J_5} - \xi_4 J_6\partial_{J_6} -$$

$$-4\xi_3 J_7\partial_{J_7} + J_8(\xi_4 - 6\xi_3)\partial_{J_8} - 2\xi_4 J_9\partial_{J_9} - (\xi_4 + 2\xi_3)J_{10}\partial_{J_{10}}.$$

Finding the first integrals of this vector field we get the basic Navier–Stokes invariants of the first order. The following result is valid for general ξ's and the special cases can be found in [13].

Theorem 6.29 ([13]) *The field of the Navier–Stokes differential invariants for the thermodynamic states admitting a one-dimensional symmetry algebra is generated by the differential invariants*

$$\frac{\xi_1}{2\xi_3 - \xi_4}\ln\rho + s, \quad J_1\rho^{\frac{\xi_4}{\xi_4 - 2\xi_3}},$$

$$\frac{J_2}{J_1}, \quad \frac{J_3}{\rho^3 J_1}, \quad \frac{J_4}{\rho J_1}, \quad \frac{J_5}{\rho^2 J_1}, \quad \frac{J_6}{J_1}, \quad \frac{J_7}{\rho^3 J_1^2}, \quad \frac{J_8}{\rho^2 J_1^2}, \quad \frac{J_9}{J_1^2}, \quad \frac{J_{10}}{\rho J_1^2}$$

of the first order and by the invariant derivations

$$\rho^{\frac{\xi_4}{\xi_4 - 2\xi_3}}\nabla_1, \quad \rho^{\frac{2\xi_3}{\xi_4 - 2\xi_3} - 1}\nabla_2, \quad \rho^{\frac{2\xi_3}{\xi_4 - 2\xi_3}}\nabla_3.$$

This field separates regular orbits.

Consider the case when the thermodynamic state admits a commutative two-dimensional symmetry algebra generated by the vector fields $A = \sum_{i=1}^{6} \mu_i X_{i+6}$, $B = \sum_{i=1}^{6} \eta_i X_{i+6}$, then μ's and η's satisfy relation

$$\eta_2 \mu_4 - \eta_4 \mu_2 = 0.$$

Performing the same computation as before we get the following theorem.

Theorem 6.30 ([13]) *The field of Navier–Stokes differential invariants for the thermodynamic states admitting a commutative two-dimensional symmetry algebra is generated by the differential invariants*

$$J_1 \rho^{\varsigma_1} e^{\varsigma_2 s}, \quad \frac{J_2}{J_1}, \quad \frac{J_3}{\rho^3 J_1}, \quad \frac{J_4}{\rho J_1}, \quad \frac{J_5}{\rho^2 J_1}, \quad \frac{J_6}{J_1}, \quad \frac{J_7}{\rho^3 J_1^2}, \quad \frac{J_8}{\rho^2 J_1^2}, \quad \frac{J_9}{J_1^2}, \quad \frac{J_{10}}{\rho J_1^2}$$

of the first order and by the invariant derivations

$$\rho^{\varsigma_1} e^{\varsigma_2 s} \nabla_1, \quad \rho^{\varsigma_1 - 2} e^{\varsigma_2 s} \nabla_2, \quad \rho^{\varsigma_1 - 1} e^{\varsigma_2 s} \nabla_3,$$

where

$$\varsigma_1 = \frac{\eta_4 \mu_1 - \eta_1 \mu_4}{2(\eta_1 \mu_3 - \eta_3 \mu_1) + \eta_4 \mu_1 - \eta_1 \mu_4}, \quad \varsigma_2 = \frac{(\eta_4 \mu_3 - \eta_3 \mu_4)}{2(\eta_1 \mu_3 - \eta_3 \mu_1) + \eta_4 \mu_1 - \eta_1 \mu_4}.$$

This field separates regular orbits.

6.4.2 3D-Flows

Consider the Navier–Stokes system (6.14) in a space $M = \mathbb{R}^3$ equipped with the coordinates (x, y, z) and the standard metric $g = dx^2 + dy^2 + dz^2$.

The velocity field of the flow has the form $\mathbf{u} = u(t, x, y, z)\,\partial_x + v(t, x, y, z)\,\partial_y + w(t, x, y, z)\,\partial_z$, the pressure p, the density ρ, the temperature T and the entropy s are the functions of time and space with the coordinates (t, x, y, z).

The vector of gravitational acceleration is of the form $\mathbf{g} = g\,\partial_z$.

6.4.2.1 Symmetry Lie Algebra

First of all we consider the Lie algebra $\mathfrak{g}$ generated by the following vector fields on the manifold $J^0 \pi$:

$$X_1 = \partial_x, \qquad X_4 = -y\,\partial_x + x\,\partial_y - v\,\partial_u + u\,\partial_v,$$

$$X_2 = \partial_y, \qquad X_5 = \left(\frac{gt^2}{2} - z\right)\partial_x + x\,\partial_z + (gt - w)\,\partial_u + u\,\partial_w,$$

$$X_3 = \partial_z, \qquad X_6 = \left(\frac{gt^2}{2} - z\right)\partial_y + y\,\partial_z + (gt - w)\,\partial_v + v\,\partial_w,$$

$$X_7 = t\,\partial_x + \partial_u, \qquad X_{10} = \partial_t,$$

$$X_8 = t\,\partial_y + \partial_v, \qquad X_{11} = \partial_s,$$

$$X_9 = t\,\partial_z + \partial_w, \qquad X_{12} = \partial_p,$$

$$X_{13} = x\,\partial_x + y\,\partial_y - \left(\frac{gt^2}{2} - z\right)\partial_z + u\,\partial_u + v\,\partial_v - (gt - w)\,\partial_w - 2\rho\,\partial_\rho + 2T\,\partial_T,$$

$$X_{14} = t\,\partial_t + gt^2\partial_z - u\,\partial_u - v\,\partial_v + (2gt - w)\,\partial_w + \rho\,\partial_\rho - p\,\partial_p - 2T\,\partial_T$$

and the Lie algebra $\mathfrak{h}$ generated by the vector fields

$$Y_1 = \partial_s, \quad Y_2 = \partial_p, \quad Y_3 = \rho\,\partial_\rho - T\,\partial_T, \quad Y_4 = p\,\partial_p + T\,\partial_T.$$

The pure geometric part is represented by the algebra $\mathfrak{g}_m = \langle X_1, X_2, \dots, X_{10}\rangle$ with respect to the group of motions, Galilean transformations, and time shifts.

In order to describe the pure thermodynamic part, we consider the Lie subalgebra $\mathfrak{h}_t$ of the algebra $\mathfrak{h}$ that preserves the thermodynamic state (6.1).

Theorem 6.31 ([14]) *A Lie algebra $\mathfrak{g}_{\mathfrak{s\eta m}}$ of symmetries of the Navier–Stokes system of differential equations in 3-dimensional space coincides with*

$$\vartheta^{-1}(\mathfrak{h}_t).$$

6.4.2.2 Symmetry Classification of States

The Lie algebra generated by the vector fields $Y_1, \dots, Y_4$ coincides with the Lie algebra of the thermodynamic symmetries of the Navier–Stokes system on a plane.

Thus the classification of the thermodynamic states or Lagrangian surfaces $\tilde{L}$ depending on the dimension of the symmetry algebra $\mathfrak{h}_t \subset \mathfrak{h}$ is the same as the classification presented in the previous section (2D-flows).

6.4.2.3 Differential Invariants

The Field of Kinematic Invariants

First of all, we observe that the functions ρ and s (as well as p and T) generate all $\mathfrak{g}_m$-invariants of order zero.

Let us consider the point $(0, \ldots, 0) \in J^0\pi$ and its isotropy group. It is easy to check that this group is isomorphic to the rotation group $SO(3)$.

Then consider the following elements:

$$
\mathbf{a_g} = \begin{pmatrix} u_t \\ v_t \\ w_t - g \end{pmatrix}, \quad \nabla\rho = \begin{pmatrix} \rho_x \\ \rho_y \\ \rho_z \end{pmatrix}, \quad \nabla s = \begin{pmatrix} s_x \\ s_y \\ s_z \end{pmatrix}, \quad V = \begin{pmatrix} u_x & u_y & u_z \\ v_x & v_y & v_z \\ w_x & w_y & w_z \end{pmatrix}
$$

and suppose that first three vectors are linearly independent.

Note that, the group $SO(3)$ acts on the matrix V by conjugacy: $V \to RVR^{-1}$, where $R \in SO(3)$.

Moreover, the action of the rotation group $SO(3)$ preserves the dot products of the vectors $\mathbf{a_g}$, $\nabla\rho$, and ∇s.

Let $H = (\mathbf{a_g}, \nabla\rho, \nabla s)$ be a matrix with $\det H \neq 0$, then the elements of the product $H^{-1}VH$ are 9 functions, which are invariant under the action of the rotation group.

Therefore, we have 15 independent invariants of the first order at the point $(0, \ldots, 0)$.

Denote by τ the following transformation:

$$
\begin{aligned}
t &\to t - t_0, & x &\to x - x_0 - u_0(t - t_0), & u &\to u - u_0, \\
\rho &\to \rho, & y &\to y - y_0 - v_0(t - t_0), & v &\to v - v_0, \\
s &\to s, & z &\to z - z_0 - w_0(t - t_0), & w &\to w - w_0.
\end{aligned}
$$

Obviously, τ is a symmetry of the equation $\mathcal{E}$, which maps the point $(t_0, x_0, y_0, z_0, u_0, v_0, w_0, \rho_0, s_0)$ to the point 0.

Applying the prolongation of τ to the invariants (to the dot products and the elements of the matrix $H^{-1}VH$) we get 15 kinematic invariants of the first order.

The proofs of the following two theorems can be found in [14].

Theorem 6.32 ([14]) *The field of the first order kinematic invariants is generated by the invariants ρ, s and by the invariants*

$$
s_t + s_x u + s_y v + s_z w,
$$

$$
(\nabla\rho)^2, \quad (\nabla s)^2, \quad \nabla\rho \cdot \nabla s, \quad (\mathbf{a_g})^2, \quad \nabla\rho \cdot \mathbf{a_g}, \quad \nabla s \cdot \mathbf{a_g}, \tag{6.17}
$$

$$
(H^{-1}VH)_{ij},
$$

transformed by τ, if $\det \mathrm{H} \neq 0$. *These invariants separate regular* $\mathfrak{g}_m$-*orbits.*

Theorem 6.33 ([14]) *The following derivations*

$$\nabla_1 = \frac{\mathrm{d}}{\mathrm{d}t} + u\frac{\mathrm{d}}{\mathrm{d}x} + v\frac{\mathrm{d}}{\mathrm{d}y} + w\frac{\mathrm{d}}{\mathrm{d}z},$$

$$\nabla_2 = \rho_x\frac{\mathrm{d}}{\mathrm{d}x} + \rho_y\frac{\mathrm{d}}{\mathrm{d}y} + \rho_z\frac{\mathrm{d}}{\mathrm{d}z}, \qquad \nabla_3 = s_x\frac{\mathrm{d}}{\mathrm{d}x} + s_y\frac{\mathrm{d}}{\mathrm{d}y} + s_z\frac{\mathrm{d}}{\mathrm{d}z},$$

$$\nabla_4 = (\rho_y s_z - \rho_z s_y)\frac{\mathrm{d}}{\mathrm{d}x} - (\rho_x s_z - \rho_z s_x)\frac{\mathrm{d}}{\mathrm{d}y} + (\rho_x s_y - \rho_y s_x)\frac{\mathrm{d}}{\mathrm{d}z}$$

are $\mathfrak{g}_m$-*invariant. They are linearly independent if*

$$\begin{vmatrix} \rho_x & \rho_y & \rho_z \\ s_x & s_y & s_z \\ \rho_y s_z - \rho_z s_y & \rho_x s_z - \rho_z s_x & \rho_x s_y - \rho_y s_x \end{vmatrix} \neq 0.$$

The bundle $\pi_{2,1} : \mathcal{E}_2 \rightarrow \mathcal{E}_1$ has rank 42 and by applying the derivations ∇_i, $i = 1, \ldots, 4$ to the kinematic invariants (6.17) we get 64 kinematic invariants. Straightforward computations show that among these invariants 42 are always independent (see https://d-omega.org).

Therefore, starting with the order $k = 1$ dimensions of regular orbits are equal to $\dim \mathfrak{g}_m = 10$.

The Hilbert function of the $\mathfrak{g}_m$-invariants field (the number of independent invariants of pure order k) is equal to

$$H(k) = \frac{9}{2}k^2 + \frac{19}{2}k + 5$$

for $k \geqslant 2$, and $H(0) = 2$, $H(1) = 16$.

The corresponding Poincaré function has the form

$$P(z) = \frac{2 + 10z - 6z^3 + 3z^4}{(1 - z)^3}.$$

Summarizing, we get the following result.

Theorem 6.34 ([14]) *The field of kinematic invariants is generated by the invariants ρ, s of order zero, by the invariants (6.17) of order one (with transformation τ), and by the invariant derivations ∇_i, $i = 1, \ldots, 4$. This field separates the regular orbits.*

The Field of Navier–Stokes Invariants

Consider the case when the equations of thermodynamic state admit a one-dimensional symmetry algebra generated by the vector field

$$A = \xi_1 X_{11} + \xi_2 X_{12} + \xi_3 X_{13} + \xi_4 X_{14}.$$

For general ξ's we have the following result. The particular cases are considered in [14].

Theorem 6.35 ([14]) *The field of the Navier–Stokes differential invariants for the thermodynamic states admitting a one-dimensional symmetry algebra is generated by the differential invariants*

$$\frac{\xi_1}{2\xi_3 - \xi_4} \ln \rho + s, \qquad \rho^{\frac{\xi_4}{\xi_4 - 2\xi_3}} \nabla_1 s, \qquad \frac{(\nabla \rho)^2}{\rho^3 \nabla_1 s}, \qquad \frac{(\nabla s)^2}{\rho \nabla_1 s}, \qquad \frac{\nabla \rho \cdot \nabla s}{\rho^2 \nabla_1 s},$$

$$\frac{\rho \,(\mathbf{a_g})^2}{(\nabla_1 s)^3}, \qquad \frac{\nabla \rho \cdot \mathbf{a_g}}{\rho \,(\nabla_1 s)^2}, \qquad \frac{\nabla s \cdot \mathbf{a_g}}{(\nabla_1 s)^2}, \qquad \frac{J_{11}}{\nabla_1 s}, \qquad \frac{J_{12}}{\rho^2}, \qquad \frac{J_{13}}{\rho},$$

$$\frac{\rho^2 J_{21}}{(\nabla_1 s)^2}, \qquad \frac{J_{22}}{\nabla_1 s}, \qquad \frac{\rho J_{23}}{\nabla_1 s}, \qquad \frac{\rho J_{31}}{(\nabla_1 s)^2}, \qquad \frac{\rho J_{32}}{\rho \nabla_1 s}, \qquad \frac{\rho J_3}{\nabla_1 s}$$

of the first order and by the invariant derivations

$$\rho^{\frac{\xi_4}{\xi_4 - 2\xi_3}} \nabla_1, \qquad \rho^{\frac{\xi_4 - 4\xi_3}{\xi_4 - 2\xi_3}} \nabla_2, \qquad \rho^{\frac{2\xi_3}{\xi_4 - 2\xi_3}} \nabla_3, \qquad \rho^{-\frac{\xi_4 - 5\xi_3}{\xi_4 - 2\xi_3}} \nabla_4,$$

here we denote by J_{ij} the elements of the matrix $\mathbf{H}^{-1}\mathbf{VH}$. This field separates the regular orbits.

Consider the case when the thermodynamic state admits a commutative two-dimensional symmetry algebra generated by the vector fields $A = \sum\limits_{i=1}^{6} \mu_i X_{i+10}$, $B = \sum\limits_{i=1}^{6} \eta_i X_{i+10}$.

Theorem 6.36 ([14]) *The field of the Navier–Stokes differential invariants for the thermodynamic states admitting a commutative two-dimensional symmetry algebra is generated by the differential invariants*

$$\rho^{\varsigma_1} e^{\varsigma_2 s} \nabla_1 s, \qquad \frac{(\nabla \rho)^2}{\rho^3 \nabla_1 s}, \qquad \frac{(\nabla s)^2}{\rho \nabla_1 s}, \qquad \frac{\nabla \rho \cdot \nabla s}{\rho^2 \nabla_1 s},$$

$$\frac{\rho (\mathbf{a_g})^2}{(\nabla_1 s)^3}, \qquad \frac{\nabla \rho \cdot \mathbf{a_g}}{\rho (\nabla_1 s)^2}, \qquad \frac{\nabla s \cdot \mathbf{a_g}}{(\nabla_1 s)^2}, \qquad \frac{J_{11}}{\nabla_1 s}, \qquad \frac{J_{12}}{\rho^2}, \qquad \frac{J_{13}}{\rho},$$

$$\frac{\rho^2 J_{21}}{(\nabla_1 s)^2}, \qquad \frac{J_{22}}{\nabla_1 s}, \qquad \frac{\rho J_{23}}{\nabla_1 s}, \qquad \frac{\rho J_{31}}{(\nabla_1 s)^2}, \qquad \frac{J_{32}}{\rho \nabla_1 s}, \qquad \frac{J_{33}}{\nabla_1 s}$$

of the first order and by the invariant derivations

$$\rho^{\varsigma_1} e^{\varsigma_2 s} \nabla_1, \qquad \rho^{\varsigma_1 - 2} e^{\varsigma_2 s} \nabla_2, \qquad \rho^{\varsigma_1 - 1} e^{\varsigma_2 s} \nabla_3, \qquad \rho^{\frac{3}{2}\varsigma_1 - \frac{5}{2}} e^{\frac{3}{2}\varsigma_2 s} \nabla_4$$

where J_{ij} are the elements of the matrix $\mathrm{H}^{-1} \mathrm{VH}$ and

$$\varsigma_1 = \frac{\eta_4 \mu_1 - \eta_1 \mu_4}{2(\eta_1 \mu_3 - \eta_3 \mu_1) + \eta_4 \mu_1 - \eta_1 \mu_4}, \qquad \varsigma_2 = \frac{2(\eta_4 \mu_3 - \eta_3 \mu_4)}{2(\eta_1 \mu_3 - \eta_3 \mu_1) + \eta_4 \mu_1 - \eta_1 \mu_4}.$$

This field separates the regular orbits.

6.4.3   Flows on a Sphere

Consider Navier–Stokes system (6.4) on a two-dimensional unit sphere $M = S^2$ with the metric $g = \sin^2 y \, dx^2 + dy^2$ in the spherical coordinates.

The velocity field of the flow has the form $\mathbf{u} = u(t, x, y) \, \partial_x + v(t, x, y) \, \partial_y$, the pressure p, the density ρ, the temperature T, and the entropy s are the functions of time and space with the coordinates (t, x, y).

Here we consider the flow without any external force field, so $\mathbf{g} = 0$.

6.4.3.1   Symmetry Lie Algebra

To describe the Lie algebra of symmetries of the Navier–Stokes system we consider the Lie algebra $\mathfrak{g}$ generated by the following vector fields on the manifold $J^0 \pi$:

$$X_1 = \partial_t, \qquad X_2 = \partial_x,$$

$$X_3 = \frac{\cos x}{\tan y}\,\partial_x + \sin x\,\partial_y - \left(\frac{\sin x}{\tan y}\,u + \frac{\cos x}{\sin^2 y}\,v\right)\partial_u + u\cos x\,\partial_v,$$

$$X_4 = \frac{\sin x}{\tan y}\,\partial_x - \cos x\,\partial_y + \left(\frac{\cos x}{\tan y}\,u - \frac{\sin x}{\sin^2 y}\,v\right)\partial_u + u\sin x\,\partial_v,$$

$$X_5 = \partial_s, \qquad X_6 = \partial_p,$$

$$X_7 = t\,\partial_t - u\,\partial_u - v\,\partial_v - p\,\partial_p + \rho\,\partial_\rho - 2T\,\partial_T,$$

and denote by $\mathfrak{h}$ the Lie algebra generated by the vector fields

$$Y_1 = \partial_s, \qquad Y_2 = \partial_p, \qquad Y_3 = p\,\partial_p - \rho\,\partial_\rho + 2T\,\partial_T.$$

Transformations corresponding to elements of the algebra $\mathfrak{g}_m = \langle X_1, X_2, X_3, X_4 \rangle$ (the pure geometric part) are generated by sphere motions and time shifts: $\mathfrak{g}_m = \mathfrak{so}(3, \mathbb{R}) \oplus \mathbb{R}$.

For describing the pure thermodynamic part we consider the Lie subalgebra $\mathfrak{h}_t$ of algebra $\mathfrak{h}$ that preserves the thermodynamic state (6.1).

Theorem 6.37 ([15]) *The Lie algebra* $\mathfrak{g}_{\mathfrak{s}\mathfrak{h}m}$ *of point symmetries of the Navier–Stokes system of differential equations on a sphere coincides with*

$$\vartheta^{-1}(\mathfrak{h}_t).$$

6.4.3.2 Symmetry Classification of States

Here we consider the thermodynamic states or Lagrangian surfaces $\tilde{L}$ (compare with the plane case) with a one-dimensional symmetry algebra $\mathfrak{h}_t \subset \mathfrak{h}$.

Cases when the thermodynamic states admit two or three-dimensional symmetry algebra are not interesting or have no physical meaning.

Let $\dim \mathfrak{h}_t = 1$ and let $Z = \sum_{i=1}^{3} \lambda_i Y_i$ be a basis vector in this algebra.

Then the thermodynamic state or the surface $\tilde{L}$ is the solution of the PDE system

$$\begin{cases} \lambda_3\rho\,\epsilon_{\rho\rho} - \lambda_1\epsilon_{\rho s} + 3\lambda_3\epsilon_\rho + \dfrac{\lambda_2}{\rho^2} = 0, \\[2mm] \lambda_1\epsilon_{ss} - \lambda_3\rho\,\epsilon_{\rho s} - 2\lambda_3\,\epsilon_s = 0, \end{cases}$$

that is formally integrable and compatible.

Solving this system for general parameters λ (some special cases can be found in [15]) we find expressions for the pressure and the temperature. Adding the admissibility conditions for this case we get the following result.

Theorem 6.38 *The thermodynamic states admitting a one-dimensional symmetry algebra have the form*

$$p = \frac{1}{\rho} \left(\frac{\lambda_1}{\lambda_3} F' - 2F \right) - \frac{\lambda_2}{\lambda_3}, \quad T = \frac{F'}{\rho^2}, \quad F = F\left(s + \frac{\lambda_1}{\lambda_3} \ln \rho \right),$$

where F is an arbitrary function and

$$F' > 0, \quad \left(\frac{\lambda_1}{\lambda_3} \right)^2 F'' - 3\frac{\lambda_1}{\lambda_3} F' + 2F > 0, \quad F'' \left(\frac{\lambda_1}{\lambda_3} F' + 2F \right) - 4(F')^2 > 0.$$

6.4.3.3 Differential Invariants

The Field of Kinematic Invariants

First of all, the pressure ρ, the entropy s, and $g(\mathbf{u}, \mathbf{u})$ (as well as p and T) generate all $\mathfrak{g}_m$-invariants of order zero.

Consider two vector fields $\mathbf{u}$ and $\tilde{\mathbf{u}}$ such that $g(\mathbf{u}, \tilde{\mathbf{u}}) = 0$ and $g(\mathbf{u}, \mathbf{u}) = g(\tilde{\mathbf{u}}, \tilde{\mathbf{u}})$. Writing the acceleration vector with respect to the vectors $\mathbf{u}$ and $\tilde{\mathbf{u}}$ we obtain two invariants of the first order. Further writing the operator $d_\nabla \mathbf{u}$ with respect to these vectors as the sum of its symmetric and antisymmetric parts we obtain another four invariants of the first order. Thus we get six invariants:

$$J_1 = (uv_t - vu_t) \sin y, \qquad J_2 = uu_t \sin^2 y + vv_t,$$

$$J_3 = u_x + v_y + v \cot y, \qquad J_4 = u_y \sin y - \frac{v_x}{\sin y} + 2u \cos y,$$

$$J_5 = (u(u_x v - v_x u) + v(u_y v - v_y u)) \sin y + u \cos y(u^2 \sin^2 y + 2v^2),$$

$$J_6 = v(u_x v - v_x u) - u(u_y v - v_y u) \sin^2 y + v^3 \cot y.$$

$$(6.18)$$

The proofs of the following theorems can be found in [15].

Theorem 6.39 ([15]) *The following derivations*

$$\nabla_1 = \frac{d}{dt}, \quad \nabla_2 = \frac{\rho_x}{\sin^2 y} \frac{d}{dx} + \rho_y \frac{d}{dy}, \quad \nabla_3 = \frac{s_x}{\sin^2 y} \frac{d}{dx} + s_y \frac{d}{dy}$$

are $\mathfrak{g}_m$-invariant. They are linearly independent if

$$\rho_x s_y - \rho_y s_x \neq 0.$$

Theorem 6.40 ([15]) *The field of the first order kinematic invariants is generated by the invariants ρ, s, $g(\mathbf{u}, \mathbf{u})$, (6.18) and*

$$\nabla_1 \rho, \quad \nabla_1 s, \quad \nabla_2 \rho, \quad \nabla_2 s, \quad \nabla_3 s. \tag{6.19}$$

These invariants separate regular $\mathfrak{g}_m$-orbits.

The bundle $\pi_{2,1} : \mathcal{E}_2 \to \mathcal{E}_1$ has rank 18, and by applying derivations $\nabla_1, \nabla_2, \nabla_3$ to the kinematic invariants (6.18) and (6.19) we get 33 kinematic invariants. Straightforward computations show that among these invariants 18 are independent (see https://d-omega.org).

Therefore, starting with the order $k = 1$ dimensions of regular orbits are equal to $\dim \mathfrak{g}_m = 4$.

Moreover, the number of independent invariants of pure order k (the Hilbert function) is equal $H(k) = 7k + 4$ for $k \geqslant 1$, and $H(0) = 3$.

The corresponding Poincaré function is

$$P(z) = \frac{3 + 5z - z^2}{(1 - z)^2}.$$

Theorem 6.41 ([15]) *The field of the kinematic invariants is generated by the invariants ρ, s, $g(\mathbf{u}, \mathbf{u})$ of order zero, by the invariants (6.18) and (6.19) of order one, and by the invariant derivations $\nabla_1, \nabla_2, \nabla_3$. This field separates regular orbits.*

The Field of Navier–Stokes Invariants

Now, we find differential invariants of the Navier–Stokes system in the case when the thermodynamic state $\tilde{L}$ admits a one-dimensional symmetry algebra generated by the vector field

$$A = \xi_1 X_5 + \xi_2 X_6 + \xi_3 X_7.$$

Theorem 6.42 ([15]) *The field of the Navier–Stokes differential invariants for thermodynamic states admitting a one-dimensional symmetry algebra is generated by the differential invariants*

$$s - \frac{\xi_1}{\xi_3} \ln \rho, \quad \rho^2 g(\mathbf{u}, \mathbf{u}), \quad \rho^3 J_1, \quad \rho^3 J_2, \quad \rho J_3, \quad \rho J_4,$$

$$\rho^3 J_5, \quad \rho^3 J_6, \quad \nabla_1 \rho, \quad \rho \nabla_1 s, \quad \frac{\nabla_2 \rho}{\rho^2}, \quad \frac{\nabla_2 s}{\rho}, \quad \nabla_3 s$$

of the first order and by the invariant derivations

$$\rho \nabla_1, \quad \rho^{-1} \nabla_2, \quad \nabla_3.$$

This field separates regular orbits.

This theorem is valid for general ξ's. For special cases see [15].

6.4.4 Flows on a Spherical Layer

Consider the Navier–Stokes system (6.14) on a spherical layer $M = S^2 \times \mathbb{R}$ with the coordinates (x, y, z) and the metric

$$g = \frac{4}{(x^2 + y^2 + 1)^2}(dx^2 + dy^2) + dz^2.$$

The velocity field of the flow has the form $\mathbf{u} = u(t, x, y, z)\,\partial_x + v(t, x, y, z)\,\partial_y + w(t, x, y, z)\,\partial_z$, the pressure p, the density ρ, the temperature T, and the entropy s are the functions of time and space with the coordinates (t, x, y, z).

The vector of gravitational acceleration is of the form $\mathbf{g} = g\,\partial_z$.

6.4.4.1 Symmetry Lie Algebra

Consider the Lie algebra $\mathfrak{g}$ generated by the following vector fields on the manifold $J^0\pi$:

$$X_1 = \partial_t, \qquad X_2 = \partial_z, \qquad X_3 = t\,\partial_z + \partial_w,$$

$$X_4 = y\,\partial_x - x\,\partial_y + v\,\partial_u - u\,\partial_v,$$

$$X_5 = xy\,\partial_x - \frac{1}{2}(x^2 - y^2 - 1)\partial_y + (xv + yu)\,\partial_u - (xu - yv)\partial_v,$$

$$X_6 = \frac{1}{2}(x^2 - y^2 + 1)\partial_x + xy\,\partial_y + (xu - yv)\partial_u + (xv + yu)\partial_v,$$

$$X_7 = \partial_s, \qquad X_8 = \partial_p,$$

$$X_9 = t\,\partial_t + gt^2\,\partial_z - u\,\partial_u - v\,\partial_v + (2gt - w)\partial_w - p\,\partial_p + \rho\,\partial_\rho - 2T\,\partial_T,$$

and denote by $\mathfrak{h}$ the Lie algebra generated by the vector fields

$$Y_1 = \partial_s, \qquad Y_2 = \partial_p, \qquad Y_3 = p\,\partial_p - \rho\,\partial_\rho + 2T\,\partial_T.$$

Transformations corresponding to the elements of the algebra $\mathfrak{g}_m = \langle X_1, \ldots, X_6 \rangle$ (the pure geometric part) are compositions of sphere motions, Galilean transformations along the z direction, and time shifts.

Let also $\mathfrak{h}_t$ be the Lie subalgebra of algebra $\mathfrak{h}$ that preserves the thermodynamic state (6.1).

Theorem 6.43 ([16]) *The Lie algebra* $\mathfrak{g}_{s\mathfrak{h}m}$ *of point symmetries of the Navier–Stokes system of differential equations on a spherical layer coincides with*

$$\vartheta^{-1}(\mathfrak{h}_t).$$

6.4.4.2   Symmetry Classification of States

The Lie algebra generated by the vector fields Y_1, Y_2, Y_3 coincides with the Lie algebra of thermodynamic symmetries of the Navier–Stokes system on a sphere.

Thus the classification of thermodynamic states or Lagrangian surfaces $\tilde{L}$ depending on the dimension of the symmetry algebra $\mathfrak{h}_t \subset \mathfrak{h}$ is the same as the classification presented in the previous section.

6.4.4.3   Differential Invariants

The Field of Kinematic Invariants

First of all, the following functions ρ, s, $g(\mathbf{u}, \mathbf{u}) - w^2$ (as well as p and T) generate all $\mathfrak{g}_m$-invariants of order zero.

The proofs of the following theorems can be found in [16].

Theorem 6.44 ([16]) *The following derivations*

$$\nabla_1 = \frac{d}{dz}, \quad \nabla_2 = \frac{d}{dt} + w\frac{d}{dz}, \quad \nabla_3 = u\frac{d}{dx} + v\frac{d}{dy}, \quad \nabla_4 = v\frac{d}{dx} - u\frac{d}{dy}$$

are $\mathfrak{g}_m$-*invariant. They are linearly independent if*

$$u^2 + v^2 \neq 0.$$

Theorem 6.45 ([16]) *The field of the first order kinematic invariants is generated by the invariants* ρ, s, $g(\mathbf{u}, \mathbf{u}) - w^2$ *of order zero and by the invariants*

$$\nabla_i \rho, \quad \nabla_i s, \quad \nabla_i(g(\mathbf{u}, \mathbf{u}) - w^2), \quad \nabla_i w,$$

$$J_1 = u_z w_x + v_z w_y, \quad J_2 = (u_x - v_y)^2 + (u_y + v_x)^2, \quad J_3 = \frac{u_t v_z - u_z v_t}{(x^2 + y^2 + 1)^2}$$

$$(6.20)$$

of order one, where $i = 1, \ldots, 4$. *These invariants separate regular* $\mathfrak{g}_m$-*orbits.*

The bundle $\pi_{2,1} : \mathcal{E}_2 \rightarrow \mathcal{E}_1$ has rank 42, and by applying derivations ∇_i, $i = 1, \ldots, 4$ to the kinematic invariants (6.20) we get 88 kinematic invariants.

Straightforward computations show that among these invariants 42 are independent (see https://d-omega.org).

Therefore, starting with the order $k = 1$ dimensions of regular orbits are equal to $\dim \mathfrak{g}_m = 6$.

The number of independent invariants of pure order k (the Hilbert function) is equal to

$$H(k) = 5 + \frac{19}{2}k + \frac{9}{2}k^2$$

for $k \geqslant 1$, and $H(0) = 3$.

The corresponding Poincaré function has the form

$$P(z) = \frac{3 + 10z - 6z^2 + 2z^3}{(1 - z)^3}.$$

Theorem 6.46 ([16]) *The field of the kinematic invariants is generated by the invariants ρ, s, $g(\mathbf{u}, \mathbf{u}) - w^2$ of order zero, by the invariants (6.20) of order one and by the invariant derivations ∇_i, $i = 1, \ldots, 4$. This field separates regular orbits.*

6.4.5 The Field of Navier–Stokes Invariants

Consider the case when the thermodynamic state $\tilde{L}$ admits a one-dimensional symmetry algebra generated by the vector field

$$A = \xi_1 X_7 + \xi_2 X_8 + \xi_3 X_9.$$

We do not consider cases of a two- or three-dimensional symmetry algebra because they are not interesting from the physical point of view.

For general ξ's we have the following theorem. For the case $\xi_3 = 0$ we have basic invariants ρ, $g(\mathbf{u}, \mathbf{u}) - w^2$, (6.20) and invariant derivatives ∇_i, $i = 1, \ldots, 4$.

Theorem 6.47 ([16]) *The field of the Navier–Stokes differential invariants for thermodynamic states admitting a one-dimensional symmetry algebra is generated by the differential invariants*

$$s - \frac{\xi_1}{\xi_3} \ln \rho, \quad \rho^2(g(\mathbf{u}, \mathbf{u}) - w^2), \quad \frac{\nabla_1 \rho}{\rho}, \quad \nabla_j \rho, \quad \nabla_1 s, \quad \rho \nabla_j s,$$

$$\rho^2 \nabla_1(g(\mathbf{u}, \mathbf{u}) - w^2), \quad \rho^3 \nabla_j(g(\mathbf{u}, \mathbf{u}) - w^2),$$

$$\rho \nabla_1 w, \quad \rho^2(\nabla_2 w - g), \quad \rho^2 \nabla_3 w, \quad \rho^2 \nabla_4 w,$$

$$\rho^2 J_1, \quad \rho^2 J_2, \quad \rho^3 J_3$$

of the first order, here $j = 2, 3, 4$, *and by the invariant derivations*

$$\nabla_1, \quad \rho\nabla_2, \quad \rho\nabla_3, \quad \rho\nabla_4.$$

This field separates regular orbits.

Acknowledgments The authors wish to express their gratitude to referees for their helpful comments and remarks on the paper.

The research was partially supported by RFBR Grant No 18-29-10013.

References

1. G. Batchelor, *An introduction to fluid dynamics*, Cambridge university press (2000).
2. J. W. Gibbs, *A Method of Geometrical Representation of the Thermodynamic Properties by Means of Surfaces*, Transactions of Connecticut Academy of Arts and Sciences, 382–404 (1873).
3. C. Carathéodory, *Untersuchungen über die Grundlagen der Thermodynamik*, Mathematische Annalen, Springer, 67, 355–386 (1909).
4. Ruppeiner, G. (1995). Riemannian geometry in thermodynamic fluctuation theory. Reviews of Modern Physics, 67(3), 605.
5. A. Bravetti, *Contact geometry and thermodynamics*, Int. J. Geom. Meth. Mod. Phys., 16, 1940003, (2018).
6. Ian M. Anderson and Charles G. Torre, *The Differential Geometry Package* (2016). Downloads. Paper 4. https://digitalcommons.usu.edu/dg_downloads/4
7. V. Lychagin, *Contact Geometry, Measurement and Thermodynamics*, in: *Nonlinear PDEs, Their Geometry and Applications. Proceedings of the Wisla 18 Summer School*, Springer Nature, Switzerland, 3–54 (2019).
8. B. Kruglikov, V. Lychagin, *Mayer brackets and solvability of PDEs–I*, Differential Geometry and its Applications, Elsevier BV, 17, 251–272 (2002).
9. B. Kruglikov, V. Lychagin, *Global Lie–Tresse theorem*, Selecta Math., 22, 1357–1411 (2016).
10. A. Duyunova, V. Lychagin, S. Tychkov, *Differential invariants for plane flows of inviscid fluids*, Analysis and Mathematical Physics, Vol. 8, No. 1, 135–154 (2018).
11. A. Duyunova, V. Lychagin, S. Tychkov, *Differential invariants for spherical flows of inviscid fluid*, Lobachevskii Journal of Mathematics, Vol. 39, No. 5, 655–663 (2018).
12. A. Duyunova, V. Lychagin, S. Tychkov, *Differential invariants for spherical layer flows of inviscid fluids*, Analysis and Mathematical Physics, doi.org/10.1007/s13324-018-0274-0 (2018).
13. A. Duyunova, V. Lychagin, S. Tychkov, *Differential invariants for plane flows of viscid fluids*, Lobachevskii Journal of Mathematics, Vol. 38, No 4, 644–652 (2017).
14. A. Duyunova, V. Lychagin, S. Tychkov, *Differential invariants for flows of viscid fluids*, Journal of Geometry and Physics, 121, 309–316 (2017).
15. A. Duyunova, V. Lychagin, S. Tychkov, *Differential invariants for spherical flows of a viscid fluid*, Journal of Geometry and Physics, 124, 436–441 (2018).
16. A. Duyunova, V. Lychagin, S. Tychkov, *Differential invariants for spherical layer flows of a viscid fluid*, Journal of Geometry and Physics, 130, 288–292 (2018).
17. M. Rosenlicht, *A remark on quotient spaces*, An. Acad. Brasil. Ci. 35, 487–489 (1963).

Printed in the United States
by Baker & Taylor Publisher Services